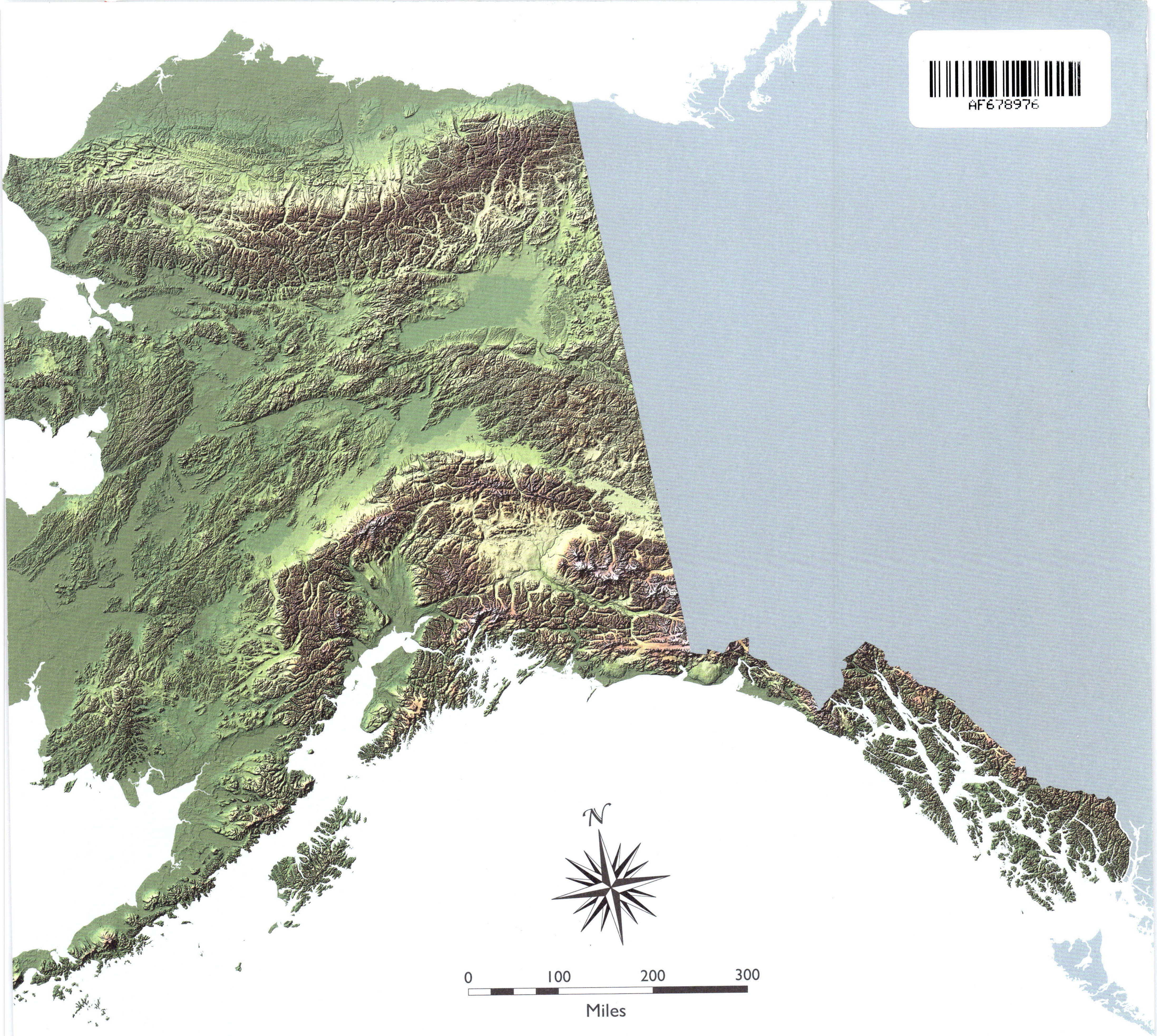
N
0
100
200
300
Miles

Sea Life of the Aleutians

An Underwater Exploration

Sea Life of the Aleutians

An Underwater Exploration

Foreword by Sylvia Earle

Reid Brewer • Héloïse Chenelot • Shawn Harper • Stephen Jewett

Alaska Sea Grant College Program
University of Alaska Fairbanks
P.O. Box 755040
Fairbanks, Alaska 99775-5040
(888) 789-0090 • (907) 474-6707
www.alaskaseagrant.org

Credits
This book, SG-ED-71, is published by the Alaska Sea Grant College Program, supported by the U.S. Department of Commerce, NOAA National Sea Grant Office, grant NA10OAR4170097, projects A/161–02 and and A/152–20; and by the University of Alaska Fairbanks with state funds.

Book design, layout, and cover design by Monica Gallo Pessino. Project management and editing by Kurt Byers and Sue Keller of Alaska Sea Grant. Cover photo by Shawn Harper; a *Eualaria fistulosa* kelp forest off Chuginidak Island, Aleutian Islands. Photo on back cover by Shawn Harper; *Hippasteria aleutica* sea star, Salmaga Pass, Aleutian Islands.

The second printing of the book is paid for by the University of Alaska Foundation, through donations from the Ted Stevens Foundation, Catherine Stevens, and the Aleut Corporation.

Elmer E. Rasmuson Library Cataloging in Publication Data:

Brewer, Reid.
Sea life of the Aleutians : an underwater exploration / Reid Brewer … [et al.]—Fairbanks, Alaska : Alaska Sea Grant College Program, 2011.
p. : ill., maps; cm. (Alaska Sea Grant College Program, University of Alaska Fairbanks; SG-ED-71)
Includes bibliographical references and index.
1. Marine biodiversity—Alaska—Aleutian Islands. 2. Marine biodiversity—North Pacific Ocean. I. Title. II. Brewer, Reid. III. Series: Alaska Sea Grant College Program, University of Alaska Fairbanks; SG-ED-71.

QH95.26.B74 2011 Second printing 2017

ISBN 978-1-56612-164-4

Printed in South Korea through Four Colour Print Group

Table of Contents

Dedication

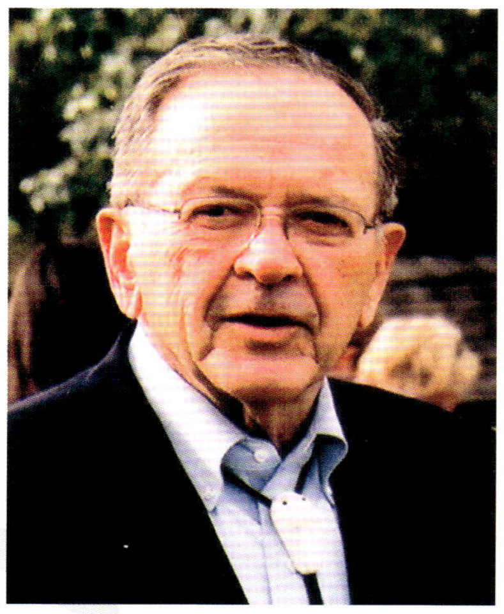

This book is dedicated to the late U.S. Senator Ted Stevens. For more than 40 years Senator Stevens served Alaska and the nation with distinction. Throughout his career he worked to enact legislation to ensure the sustainability of marine resources in Alaska and nationwide, and to strengthen Alaska communities.

In 1976, Senator Stevens coauthored the Fishery Conservation and Management Act, which established the 200-mile Exclusive Economic Zone and wrested control of the zone's rich marine resources from unregulated foreign fishing fleets. Later renamed in honor of its cosponsors, Senator Stevens and Washington Senator Warren Magnuson, the Magnuson-Stevens Fishery Conservation and Management Act "Americanized" fisheries worth billions of dollars and established a science-based, public process to manage U.S. marine resources for long-term sustainability. Application of the Magnuson-Stevens Act in Alaska waters—where more than half the nation's annual seafood is landed—is considered a model of effective, responsible fishery management.

In subsequent years, Stevens authored amendments to the Magnuson-Stevens Act to address new concerns. This included restructuring the Seattle-based Bering Sea pollock fishery, the nation's largest fishery, to ensure its long-term economic and biological viability. Stevens crafted an amendment stipulating that pollock fishing companies share a portion of their lucrative Bering Sea harvest with small villages along the western Alaska coast, which border the fishing grounds but did not benefit from the ocean bounty off their shores.

Recognizing the importance of science as the foundation for sound resource management decisions, Senator Stevens focused national resources on marine research and education. He championed the National Sea Grant College Program, the Alaska SeaLife Center, and the Alaska Region Research Vessel *Sikuliaq*, and he endowed the North Pacific Research Board to ensure a robust and diverse marine research effort in the North Pacific and Arctic oceans.

Senator Stevens was concerned about the socioeconomic implications of Steller sea lion population declines in western Alaska. He procured significant funds for Steller sea lion research in Alaska, which enabled state and federal management agencies, the Alaska SeaLife Center, and the University of Alaska Fairbanks to examine possible causes of those declines.

Highlighting his keen interest in Alaska commercial fishing and support for Alaska Sea Grant's educational efforts, Senator Stevens wrote the foreword to Alaska Sea Grant's book, *Ocean Treasure: Commercial Fishing in Alaska* (2003). In it he stated, "I am dedicated to exploring measures that will improve the management of our fisheries. My interest in this industry extends beyond policy; as a recreational fisherman and father of a former commercial fisherman, my goal is to ensure that this resource is available to future generations. Fishing is Alaska's greatest pastime." Senator Stevens passed away pursuing that pastime, in an aviation accident August 9, 2010, while on a fishing excursion in western Alaska.

His legacy in marine science and education is honored with the Ted Stevens Distinguished Professorship in Marine Policy at the University of Alaska Fairbanks School of Fisheries and Ocean Sciences. In 2007 the National Oceanic and Atmospheric Administration opened Alaska's largest fisheries research facility, the Ted Stevens Marine Research Institute, at Lena Point near Juneau.

Senator Ted Stevens played a pivotal role in Alaska history—in establishing statehood, building the Trans Alaska Pipeline, resolving Alaska Native land claims, strengthening the nation's defense capabilities in Alaska, and investing statewide in roads, harbors, airports, hospitals, schools, and other community needs essential to our quality of life. For his tireless efforts on behalf of all Alaskans, and particularly for his dedication to Alaska fisheries and marine science, we dedicate this book to the late U.S. Senator Ted Stevens.

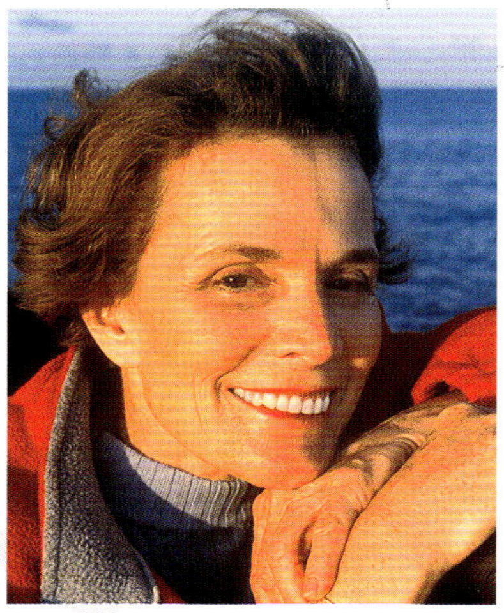

Alaska's Wild Ocean

Foreword by Sylvia Earle

For at least nine millennia, people have thrived along the twelve-hundred-mile arc of jagged Aleutian Islands that span the ocean between western Alaska and Russia's Far East. Preceding them as residents by millions of years are the amazingly rich and diverse kinds of creatures celebrated in this volume.

Beautifully illustrated here are a dozen of the 30 or so major divisions of animal forms that occupy the earth, from sponges and comb jellies to anemones, arthropods, echinoderms, mollusks, lamp shells, polychaetes, phoronids, bryozoans, sea squirts, and variations on the theme of humankind's fellow vertebrates, the fishes. At least another dozen less conspicuous divisions of life are also likely to be in these images, woven as a living tapestry over rocks, on blades of kelp, within the sponges, the tissues of corals, and in the surrounding water column, itself a thriving minestrone of creatures of ancient lineage. Just as in tropical coral reefs, the waters of the Aleutian Islands are home for species representing most of the major divisions of the earth's animals, plants, and microbes. Only about half of these great wedges of life occur above the sea.

By the beginning of the 20th century, many of the islands' birds, otters, seals, fish, and whales had significantly declined, at first captured by hunters to support basic sustenance, but in time depleted to supply the demands of distant markets. The last of the giant manatees, Steller's sea cow, was killed in 1768, just 27 years after discovery by Europeans during an expedition led by Vitus Bering. Over the years, millions of tons of sea life have been taken from these stunningly productive waters with increasingly sophisticated means of exploitation.

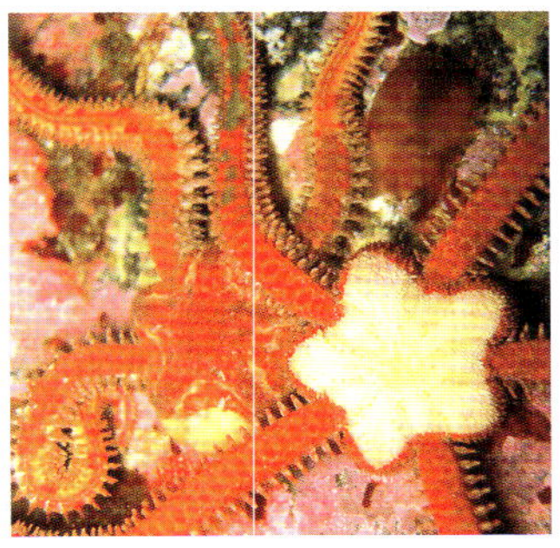 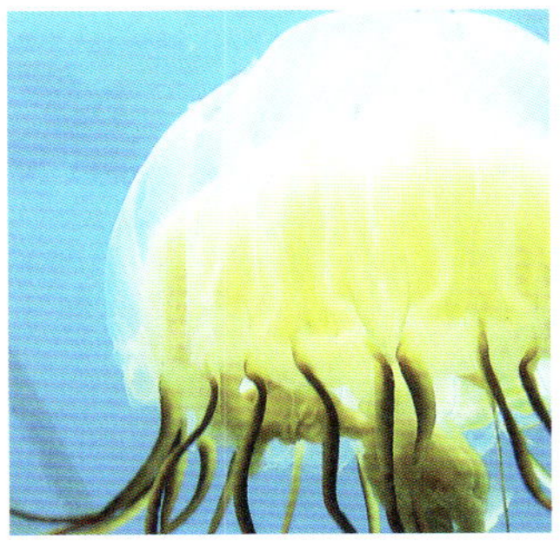 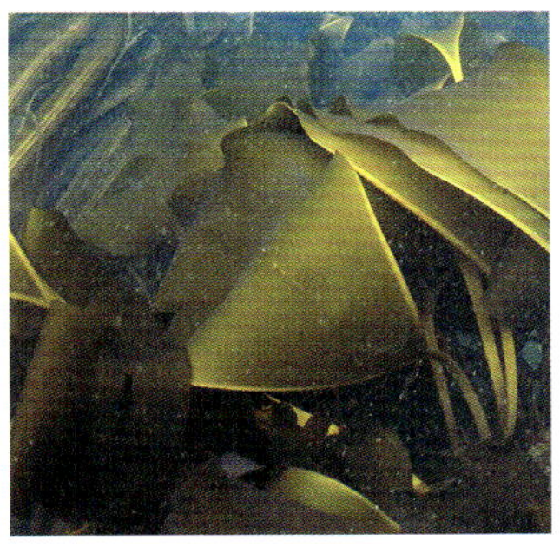 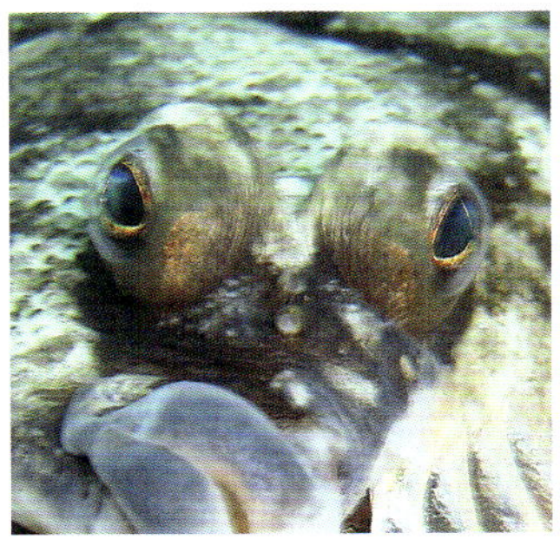 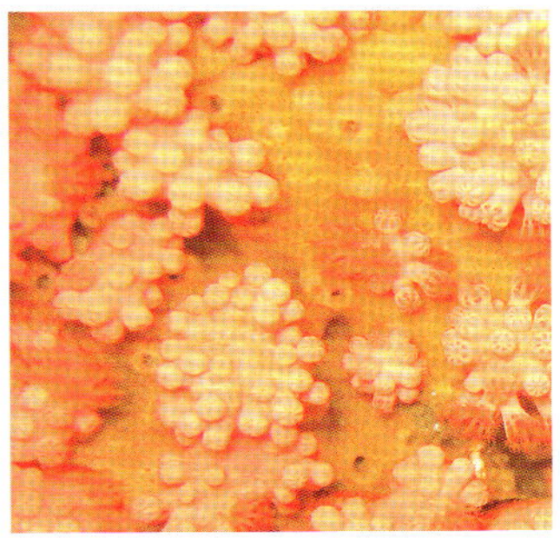

Exploration, meanwhile, has awaited the development of suitable technologies—and the will and means to apply them. Diving with scuba, an approach used effectively by scientists and explorers since the mid 1950s, requires a special kind of commitment in the Aleutians. The water is typically clear but always cold; storms are common and impenetrable veils of fog regularly obscure sky, land, and sea. Adventures of crab fishermen depicted in a popular television series suggest that no sane person would choose to deliberately venture beneath the surface.

Yet, Ted Stevens, to whom this book is dedicated, wished to do so. He once confided to me that as a senator-to-be teenager, he was motivated to create his personal version of a diving system based on a modified water heater, long before the availability of regulators, tanks, dry suits, underwater cameras, GPS units, computers, satellite phones, and other gear used by the authors of this volume. Thanks to these intrepid individuals, anyone who wishes to experience the world of living treasures that grace the depths along the Aleutian Islands can vicariously do so.

Knowledge that forests and beds of green, brown, and red algae and a diverse menagerie of creatures border these islands in depths to at least 20 meters should inspire measures to protect them—and with precautionary wisdom, extend care to the still largely unexplored deep sea beyond. Now we know what our ancient predecessors could not: the living ocean is the life support system for all of life on earth, ourselves very much included.

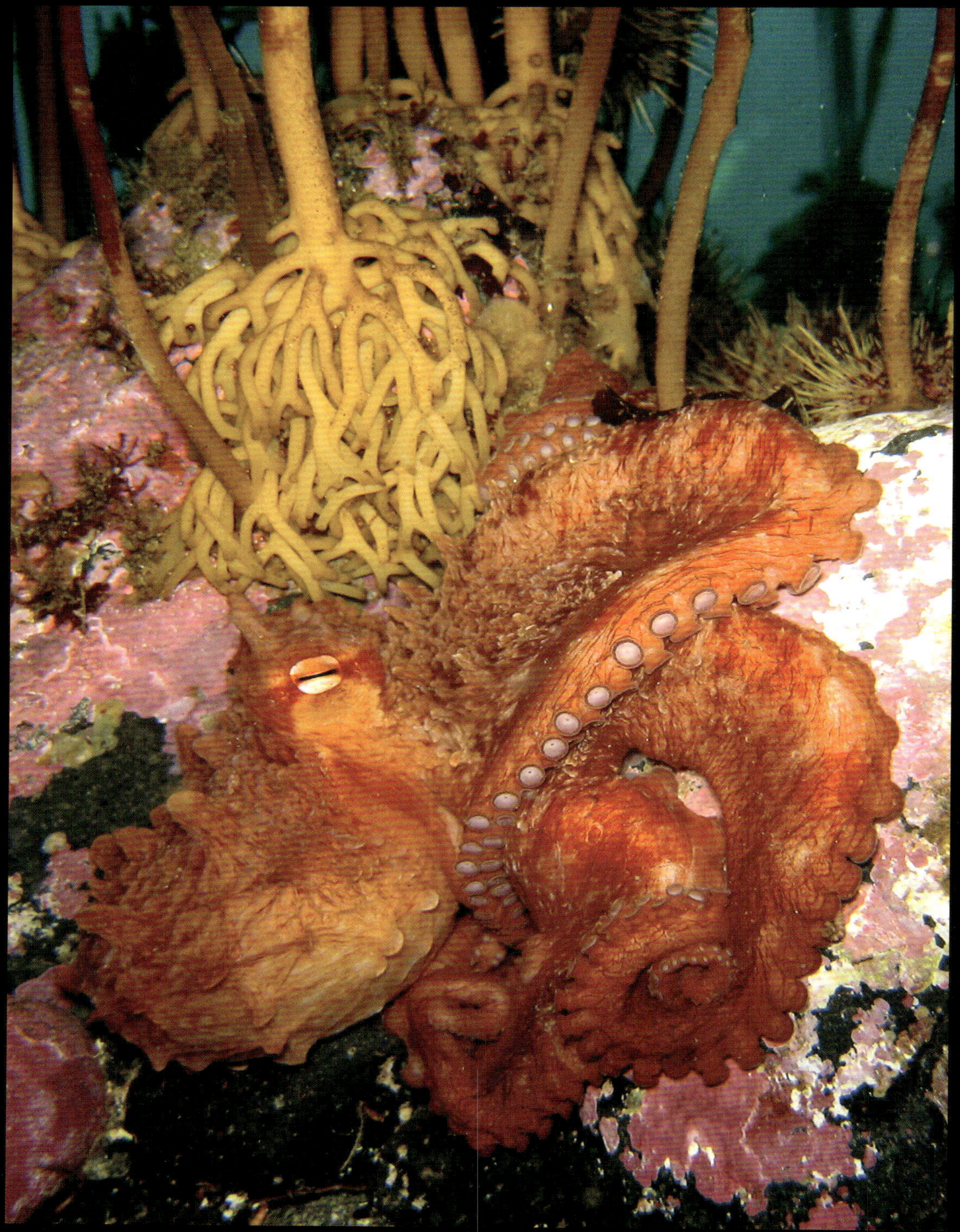

The Aleutian Islands truly are an octopus's garden. This North Pacific giant octopus (*Enteroctopus dofleini*) hides at the base of a kelp near Adak Island. *Kagalaska Island, R. Brewer photo.*

Introduction

If only Georg Wilhelm Steller had looked down—or better, could have scuba dived.

Of course, scuba diving was still some 200 years in the future when in 1741 the German naturalist joined Danish sea captain, Vitus Jonassen Bering, on a far-ranging Russia-financed ocean voyage, an expedition that resulted in discovery of some of the Aleutian Islands. Steller probably wasn't as interested in the creatures that lived beneath the frigid waves of the Bering Sea as he was in the animals and plants that teemed across the region's 1,200-mile daisy chain of windswept islands.

Steller would lend his name to many of the new animals he saw—the Steller sea lion, Steller's eagle, Steller's jay, Steller's eider, and the soon thereafter extinct manatee-like Steller's sea cow. Discoveries of the colorful and exotic marine life beneath the rolling Aleutian waves would have to wait.

It would be a long wait. In the 270 years after Steller became the first European to set foot on what is now Alaska, much has happened in the Aleutians. The Russian and American sea otter and fur seal trade that followed in the wake of Bering's ships—the *Sviatoi Piotr* (St. Peter) and *Sviatoi Pavel* (St. Paul)—decimated not just the animals but also the Native Aleut people. Yankee whalers exacted a massive toll on right, sperm, minke, humpback, and other whales.

World War II saw thousands of American GIs battling both the enemy and hostile elements for control of far-flung islands with curious names like Attu and Adak. The Cold War that came next brought underground testing of nuclear weapons far more powerful than the ones dropped on Hiroshima and Nagasaki—including the largest atomic weapon ever detonated in the United States, a 5-megaton blast at Amchitka in 1971, the force of which registered 7 on the Richter Scale.

As all of these events unfolded, the undersea life of the Aleutians carried on pretty much unnoticed and undocumented until now.

In 2004, 2006, and 2007, researchers led by marine scientists from the University of Alaska Fairbanks embarked on missions to explore and document the undersea life of the Aleutian Islands. Three years and 676 dives later, these intrepid explorers performed the first Aleutian-wide study of nearshore sea life.

Dr. Stephen Jewett, lead scientist and coauthor, describes his first dive in the Aleutians.

> As I rolled off the dive boat and glanced toward the bottom I was amazed at the water clarity. Visibility in most places where I dive in Alaska usually is less than 15 feet. But on this 20-foot dive I could see out to 100 feet. Conditions were optimal for such incredible visibility—calm, little suspended sediment and plankton, and virtually no current.
>
> The light-pink encrusting coralline algae that covered much of the rocky bottom created a bright surrounding to the seabed, illuminating an incredible diversity of life. As I slowly drifted toward the bottom I was awestruck at what lay below—a small forest of kelp that reached to the sea surface, and a menagerie of brilliantly colored plants and animals.
>
> As I approached the seafloor, the colorful mosaic consisted of bright orange vase sponges nearly three feet tall; numerous species of sea anemones in shades of white, pink, green, and orange; lavender soft corals; orange sea squirts; patches of white, orange-red, and yellow sponges and sea urchins; and several species of sea stars. An occasional male rock greenling, with its mottled vermillion pattern, and black rockfish appeared like sentinels guarding this treasured environment. There were entire colonies of ascidians, hydroids, and bryozoans as well as thousands of brittle stars—their arms extending from the crevices of the coralline algae. This seemed to be a marine Garden of Eden.

Sea Life of the Aleutians: An Underwater Exploration is a mere sampling of the 5,000-plus photographs taken of the marine organisms encountered on these trips.

The intent of this book is to share the beauty and diversity of the nearshore marine life of the Aleutian Islands with the rest of the world and encourage future generations to continue to explore and conserve the amazing wonders under the sea.

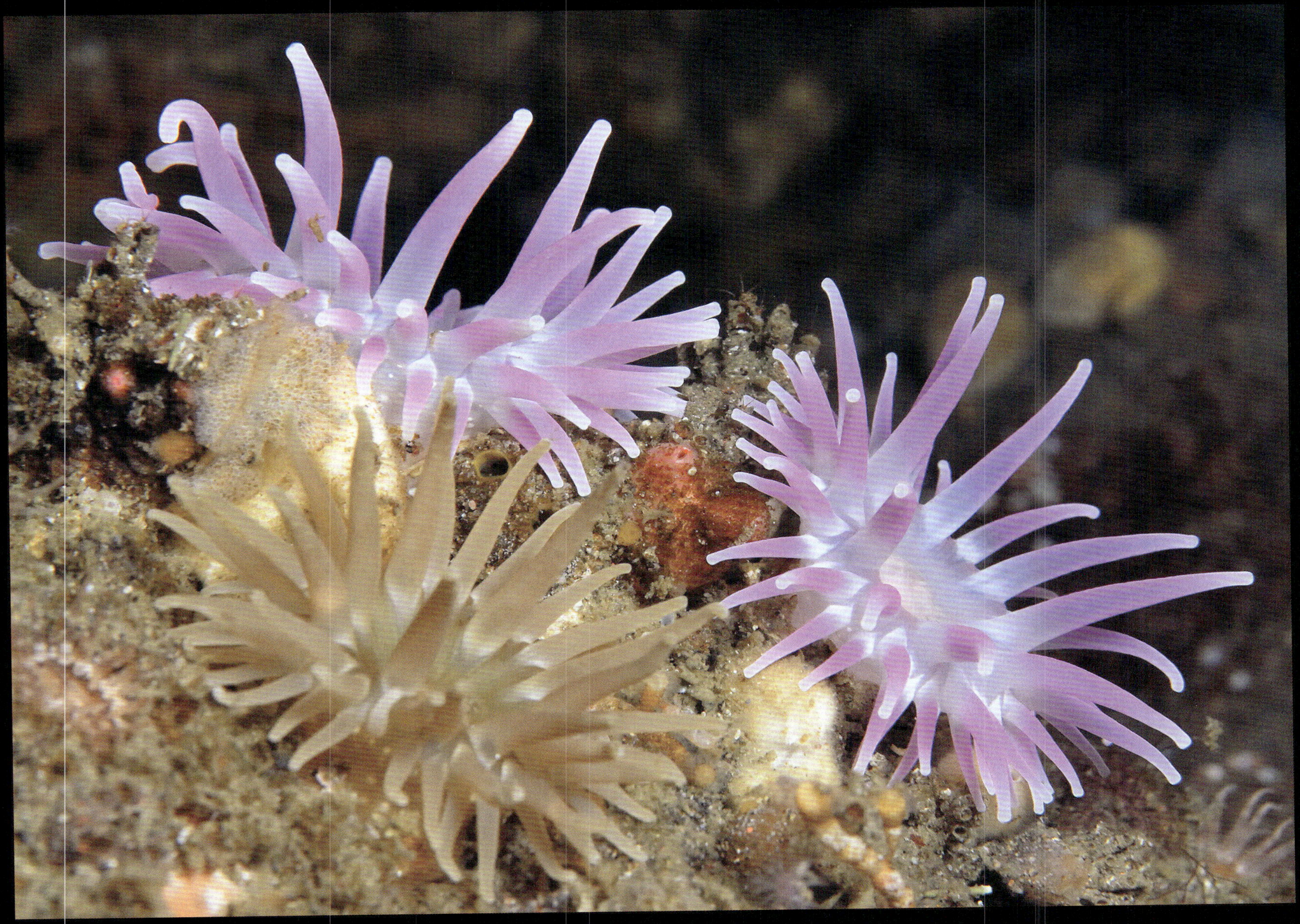

Colorful sea anemones often live near hydrothermal vents in the Islands of Four Mountains. These anemones inhabit high flow areas where food is delivered to their doorstep. *Kagamil Pass, Kagamil Island, S. Harper photo.*

The Setting

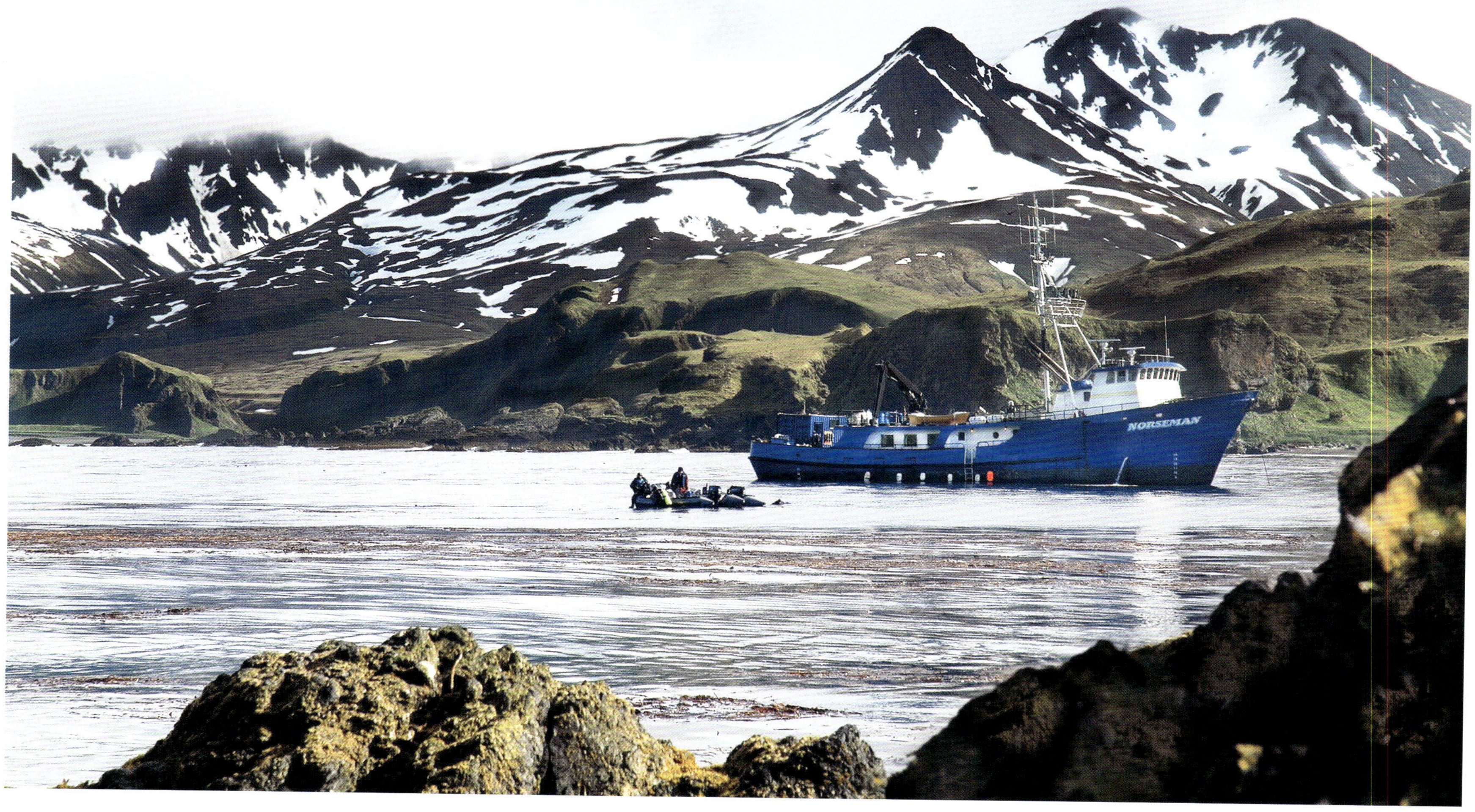

The 108-foot (33 m) converted crabber, R/V *Norseman*, stands by while divers prepare to sample between Savage Island and Attu Island at the western tip of the Aleutian archipelago, June 2007. *M. Lindeberg photo.*

The Research Expeditions

The photographs in this book were mainly taken during three research cruises, in 2004, 2006, and 2007. A few additional photographs were taken during a 2008 cruise in the western Aleutians. In 2004, a team of scientists dived around Amchitka and Kiska islands in the west-central Aleutians to sample the nearshore benthic community for radionuclides, possibly leaking from nuclear detonations of the U.S. Atomic Energy Commission on Amchitka Island in 1965, 1969, and 1971. No radioactivity was found. In 2006 and 2007 a team of research divers randomly sampled 50 sites in the nearshore Aleutian Islands in a contaminant-based investigation that was funded by the U.S. Environmental Protection Agency. All photos were taken during June and July at depths generally less than 60 feet (20 meters).

Diversity and Discovery

The nearshore Aleutian Islands are unique because of extraordinary underwater visibility, often approaching 100 feet (30 meters). As a result of limited sediment output from rivers and light penetration into upwelled, nutrient-rich waters, biodiversity in the Aleutians is high.

Biodiversity in coastal regions, beyond coral reefs, has received increasing attention in recent years. The three-dimensional structure of macroalgal habitats and the sediment turnover in seagrass communities allow for a stable yet diverse ecosystem. Unfortunately, these same habitats are most frequently impacted by human activities. Pollution, fishing, and other recreational, industrial, and military activities can severely impact nearshore biodiversity. On larger scales, global climate change may have a significant impact on the environment and thus on marine ecosystems.

This book highlights the high biodiversity of the nearshore marine community of the Aleutian Islands. At many of the dive sites, particularly in kelp forests, diversity is exceptional with several dozen species identified within the limits of small (1 x 30 m) transects. More than 1,000 species of algae, invertebrates, and fishes were identified from the 50 sites sampled in 2006 and 2007. Many taxa were not identified to the species level, but several new species are emerging as taxonomic analyses continue. As of 2011, the following discoveries were confirmed:

- 52 range extensions of benthic marine algae (work in progress)
- 15 new species of benthic marine algae (work in progress)
- 1 new genus and species of benthic marine algae (published)
- 9 of the new species of benthic marine algae are endemic to the Aleutian Islands
- 1 new species of sea anemone (published)
- 3 new species of gastropods (work in progress)
- 1 new genus of chiton (work in progress)
- 5 new species of chitons (work in progress)
- 1 new genus of sea star (published)
- 17 new species of sea stars (published)

Some species included in this book are found throughout the North Pacific, others occur only in Alaska, and a small number are endemic to the Aleutian Islands. On an even smaller scale, some species appear to be limited to distinct Aleutian Island groups. Algal and faunal breaks within the Aleutians are attributed to the numerous deepwater passes that act as barriers to the distribution of some species. The identities of plants and animals shown in this book were determined through group consensus by examining samples and photographs and through consultation with taxonomic experts.

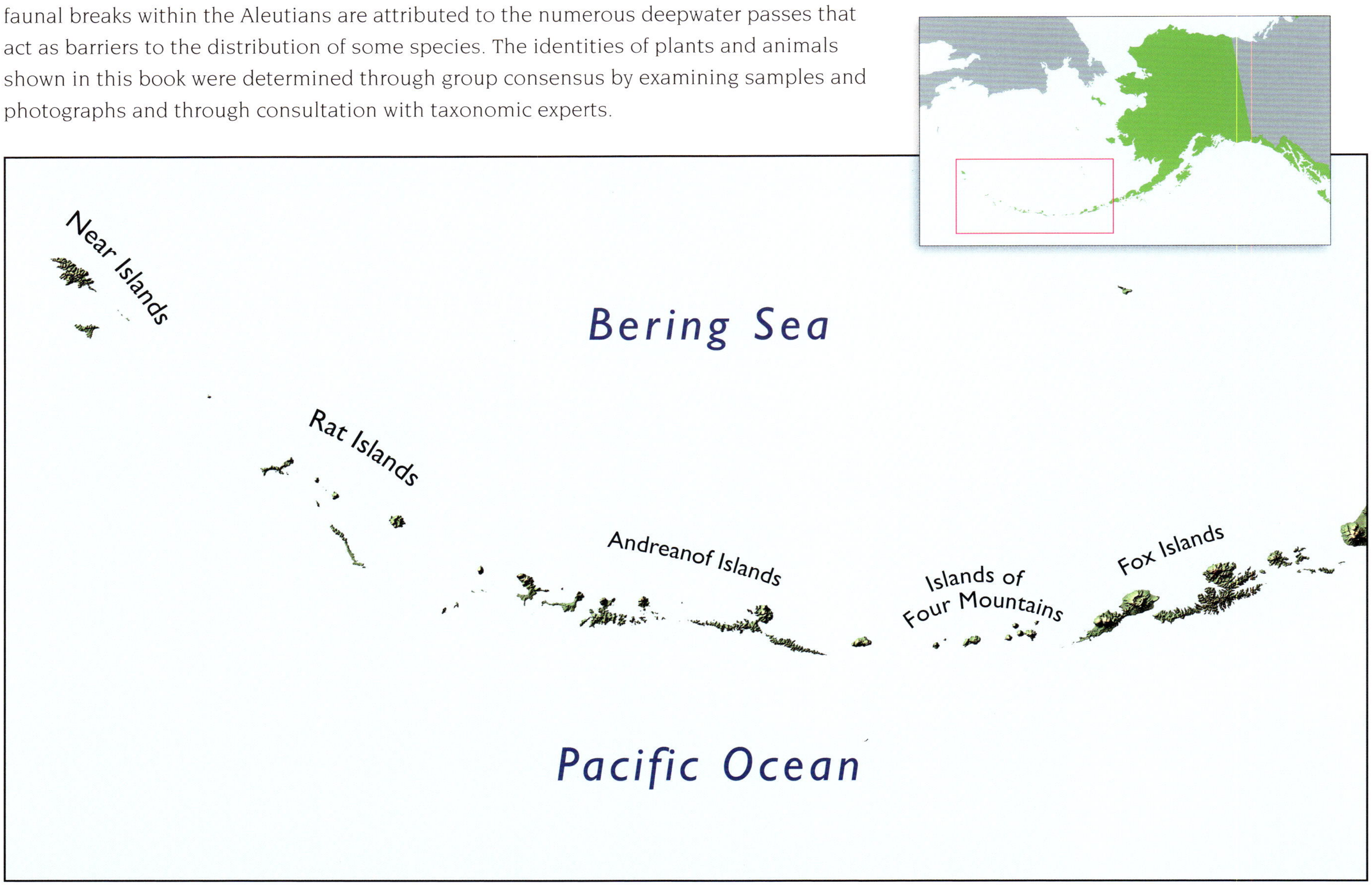

Geography

The Aleutian archipelago lies west of the Alaska Peninsula, extending westward from Unimak Island to Attu Island, a distance of nearly 1,200 miles (1,931 km). The Aleutian archipelago comprises five main island groups and over 200 islands totaling about 2.7 million acres (110,000 km^2). This vast expanse forms an arc that separates the North Pacific Ocean from the Bering Sea.

The central area of the Aleutian chain is the southernmost part of Alaska, dipping below 52°N latitude, and because the 180° longitude line runs north-south through Amchitka Pass, the Aleutians include both the easternmost and westernmost points in the United States.

The Fox Islands lie between the Alaska Peninsula and just east of Samalga Pass and the Islands of Four Mountains. The largest islands in this group include Akutan, Unalaska, and Umnak. The small island group of the Krenitzen Islands is a subset of the Fox Islands. Tigalda Island was the most densely populated of the Fox Islands before Westerners arrived. When Vitus Bering first came to these islands, he found indigenous red foxes and thus named them the Fox Islands.

The Islands of Four Mountains are located between Amukta Pass and the Andreanof Islands to the west, and Samalga Pass and the Fox Islands to the east. This island group includes Kagamil, Chuginadak, Carlisle, Herbert, Yunaska, and Amukta. The name of this island group was applied by the early Russian explorers because of four prominent volcanoes located on four of the islands: Kagamil, Carlisle, Chuginadak, and Amukta.

The Andreanof Islands are the largest of the Aleutian Island groups. They are located between Amchitka Pass and the Rat Islands group to the west, and Amukta Pass and the Islands of Four Mountains to the east. The Andreanof Islands range from Seguam to Gareloi, encompassing Amlia, Atka, Great Sitkin, Adak, Kanaga, and Tanaga. The Delarof Islands are a subset of the Andreanof Islands. The Andreanof Islands were named for the Russian navigator, Andrean Tolstyk, who was the first to explore the islands in 1761.

The Rat Islands are located between Buldir Island and the Near Islands group to the west, and Amchitka Pass and the Andreanof Islands to the east. The largest islands of the Rat Islands are Semisopochnoi, Amchitka, Little Sitkin, and Kiska. The name was applied to the islands because rats ruled Rat Island from about 1780 on. In 2009, the U.S. Fish and Wildlife Service completed an extensive rat eradication program on Rat Island to enable bird populations to recolonize the island.

The Near Islands are the smallest and westernmost island group of the Aleutians. They include Agattu and Attu. A smaller island group known as the Semichi Islands, which includes Shemya, is a subset of the Near Islands group. They were named Near Islands by Russian explorers in the 1700s because they were the nearest of the Aleutian Islands to Russia.

West of the Near Islands, the Commander Islands, part of Russia, continue the island arc and are biologically linked to the Kamchatka shelf and coast. South of this submerged mountain range, of which the Aleutian Islands are the exposed peaks, is a curving submarine trench as deep as 25,000 feet (7,600 m). The trench extends across the North Pacific for 2,000 miles (3,200 km) from the Gulf of Alaska to the Kamchatka Peninsula.

Volcanology

The Aleutian Islands rose from the volcanic activity resulting from the convergence of the Pacific and North American plates. Today the Aleutian Islands remain one of the most seismically and volcanically active regions in the world. Part of the Pacific Ring of Fire, the Aleutians have 57 volcanoes, 27 of which are active. The tallest volcano in the Aleutians, Shishaldin Volcano on Unimak Island, reaches 9,372 ft (2,856 m). Thirteen other volcanoes are taller than 5,000 ft (1,524 m).

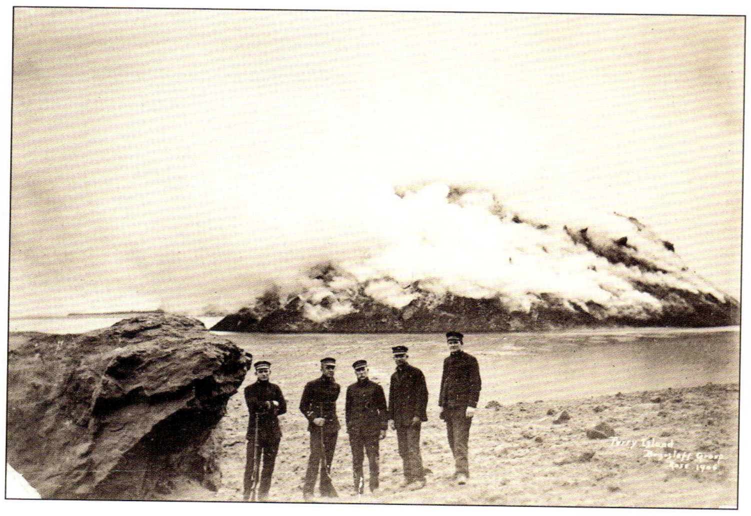

U.S. Revenue Cutter Service sailors stand on Bogoslof Island as Perry Island wells up from the sea in 1906. Bogoslof Island was formed by volcanic action in 1796. *Courtesy UAF Alaska and Polar Regions Collections.*

Active volcanoes are still shaping many of the islands today, altering old habitats and creating new ones. Perry Island, adjacent to Bogoslof Island, rose above sea level just over a hundred years ago. In August 2008, a volcano on Kasatochi Island in the Andreanof Islands erupted, changing the face of Kasatochi Island forever (see sidebar).

Kasatochi Volcano

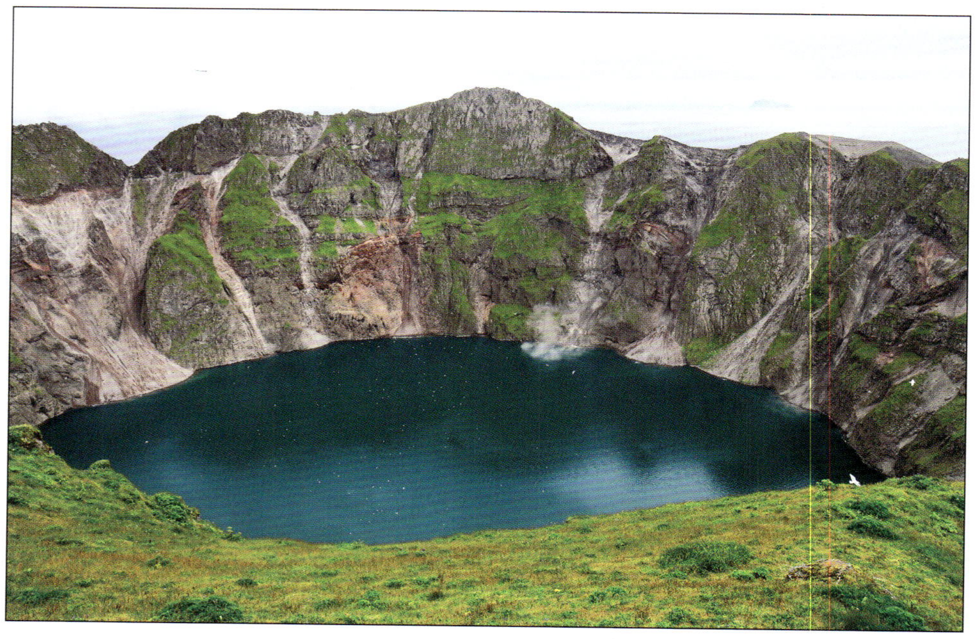

Kasatochi crater, August 6, 2008, the day before the dormant volcano erupted. *C. Ford photo, USFWS.*

Kasatochi Volcano in Alaska's Aleutian Islands suddenly erupted on August 7, 2008, completely covering the island and intertidal and subtidal zones with ash and volcanic material. Two U.S. Fish and Wildlife Service (USFWS) scientists stationed on the island narrowly escaped death, rescued by a commercial fisherman just before the volcano blew. Before the blast, Kasatochi was lush and green and boasted a beautiful blue lake in a deep caldera. Birds—including 100,000 crested and least auklets, storm petrels, and puffins—lived on the 700-acre island.

The eruption enlarged the diameter of the summit crater by 100 meters and extended the shoreline outward from the island about 400 meters. Thousands of seabird chicks were entombed under the ash. The eruption also killed all plants, fish, land animals, and most insects, virtually sterilizing the small isle. The thick ash smothered the nearshore underwater kelp forest habitat and all of the organisms that made their homes there. University and government scientists are taking advantage of this rare opportunity to document how biota will recolonize the island and its nearshore underwater habitat. —*Adapted in part from reports by U.S. Geological Survey, University of Alaska Museum, and Alaska Science Forum.*

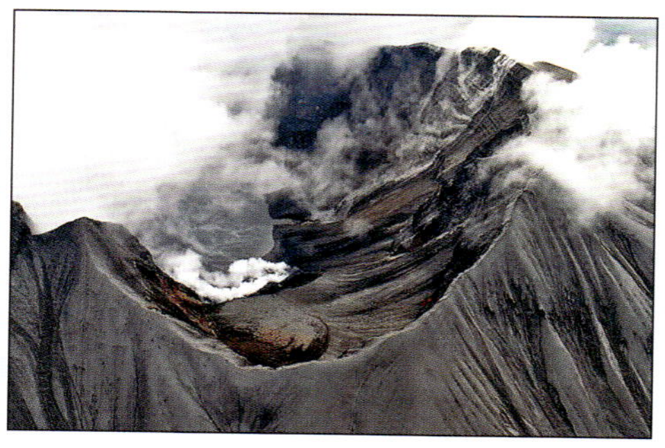

Kasatochi caldera two weeks after the eruption. *C. Waythomas photo, Alaska Volcano Observatory/USGS.*

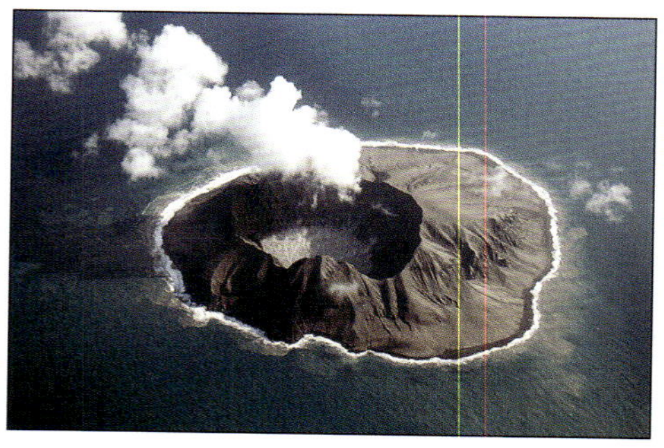

Kasatochi Volcano, October 23, 2008, three months after it erupted. *J. Morris photo.*

Ocean Currents

The strong Alaska Coastal Current flows west along the south edge of the Aleutians, and the Aleutian North Slope Current moves east toward the Alaska mainland on the north side of the Aleutians.

A large volume of water from the Alaska Coastal Current moves through 14 island passes in the Aleutians to the Aleutian North Slope Current, providing relatively fresh surface waters and warm subsurface waters to the Bering Sea on the north side of the island chain. The Aleutian North Slope Current moves up the steep rise of the continental slope in the Bering Sea and directs much of the water into a northwest-flowing Bering Slope Current. The upwelling of nutrient-rich water through the Aleutians and onto the continental shelf results in a highly productive marine system.

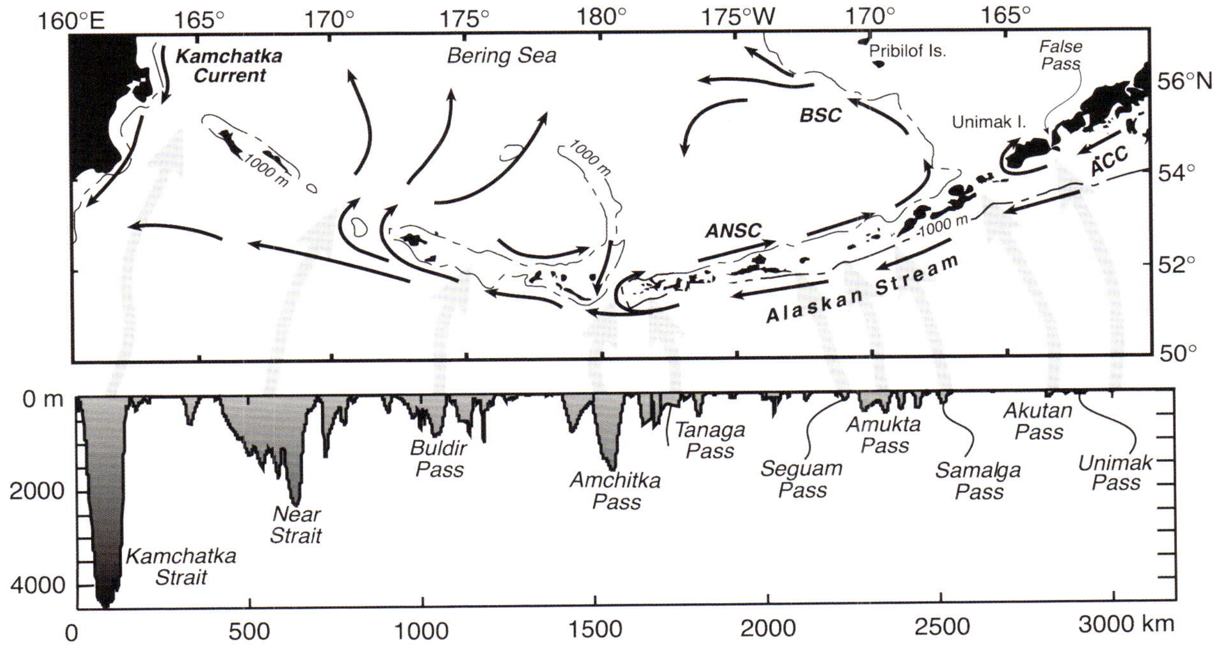

Resource Management

Most of the Aleutian Islands are part of the Alaska Maritime National Wildlife Refuge, managed by the U.S. Fish and Wildlife Service. State and federal fisheries managers have established many marine protected areas in the waters off of Alaska, including the Aleutians. These management efforts have been put into place to protect ecological structure and function, establish control sites for scientific research studies, conserve benthic habitat, protect vulnerable stocks, and protect cultural resources. Alaska Native groups also help manage use of natural resources in the Aleutians.

To address concerns about the impacts of bottom trawling on benthic habitat, particularly on coral and sponge communities in the Aleutian Islands, the North Pacific Fishery Management Council and the National Marine Fisheries Service took action to prohibit all bottom trawling, except in small discrete "open" areas. Beginning in 2006, over 95 percent of the Aleutian Islands management area was closed to bottom trawling, approximately 277,100 square nautical miles (950,463 km^2).

Near the end of the twentieth century, large-scale ecosystem impacts were documented in the Aleutian Island and Bering Sea region, which have been attributed to climate change and human impacts on the system. Effects of numerous anthropogenic stressors, ranging from contaminants to commercial fisheries to invasive species, need to be understood if resource managers are to develop and practice adaptive management.

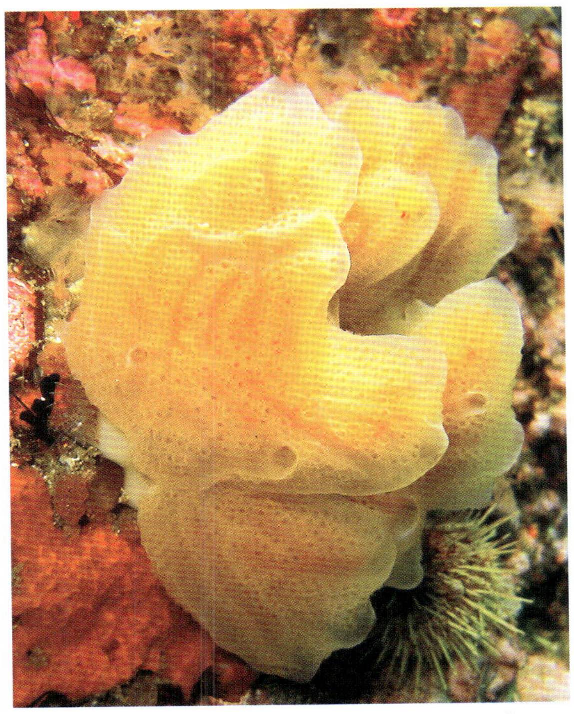

Ascidians, also known as sea squirts, attach to hard surfaces where they filter-feed. *Amchitka Island, H. Chenelot photo.*

About the authors

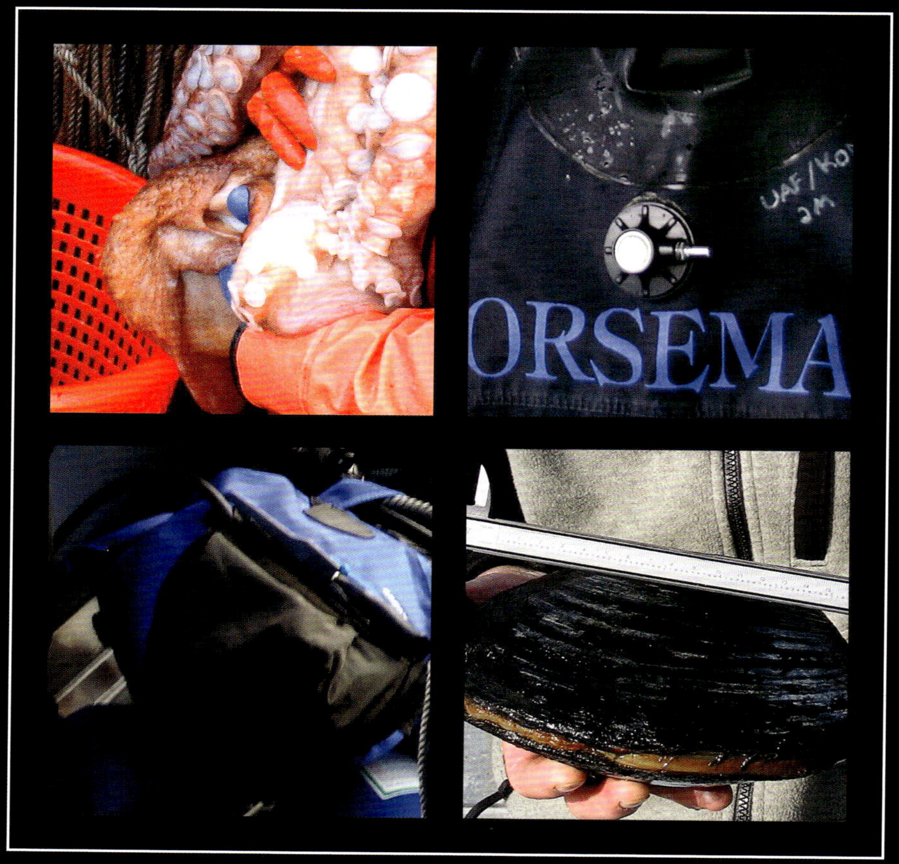

Reid Brewer

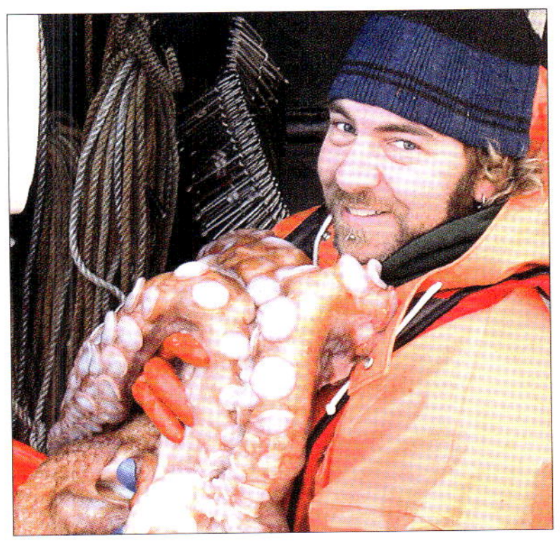

Reid Brewer has been an Alaska Sea Grant Marine Advisory Program agent, based at Unalaska in the Aleutians, since 2004. He is an associate professor with the University of Alaska Fairbanks (UAF) School of Fisheries and Ocean Sciences, and at the time of this printing he is working on his doctoral degree at UAF in marine biology. He is a PADI Master Scuba Diver Trainer, and has certified more than 100 scuba divers in Unalaska. He founded and chairs the Unalaska Divers Association, and has made over 1,000 Alaska dives. An alum of the U.S. Military Academy at West Point and former U.S. Army helicopter pilot and Rappel Master, Reid devotes most of his time to education, outreach, research, and service to communities in the Aleutians, and his greatest passion is scientific diving. He has participated in and led numerous invertebrate projects throughout the Aleutians and is an authority on the natural history of the North Pacific giant octopus. His underwater photos have appeared in *Aurora* and *Diver* magazines, *Field Guide to Seaweeds of Alaska*, *Dutch Harbor Fisherman*, and *Anchorage Daily News* and were featured in a gallery exhibit at the Museum of the Aleutians. *Zach Nehus photo.*

Héloïse Chenelot

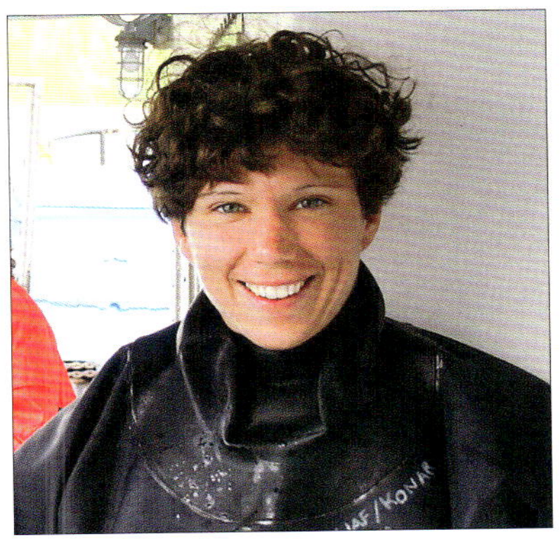

Héloïse Chenelot is a research technician at the University of Alaska Fairbanks (UAF) School of Fisheries and Ocean Sciences. She earned her master's degree in biological oceanography from UAF in 2003. Most of her diving experience has been logged in Alaska, from Kachemak Bay to the Aleutians and the Arctic. Héloïse's primary research interest is biodiversity in nearshore environments. She feels privileged to be involved in many scientific projects that explore the unique and beautiful marine life of Alaska.
While Héloïse knows that nothing can replace the awe-inspiring feeling of experiencing in person the spectacular beauty of the underwater world, she believes that photographs are a powerful means for educating others about science and instilling the need to preserve nature's splendor. Héloïse's underwater photos have appeared in the *Alaska Coastal Calendar*, back cover of *Field Guide to Common Marine Fishes and Invertebrates of Alaska*, in the books *Imam Cimiucia: Our Changing Sea* and *Field Guide to Seaweeds of Alaska*, on the front cover and in the Pollock Conservation Cooperative Research Center research report, on the cover of the 2011 Alaska Sea Grant catalog, in *Aurora* magazine, and at the Museum of the Aleutians. *Paul Tate photo.*

Shawn Harper

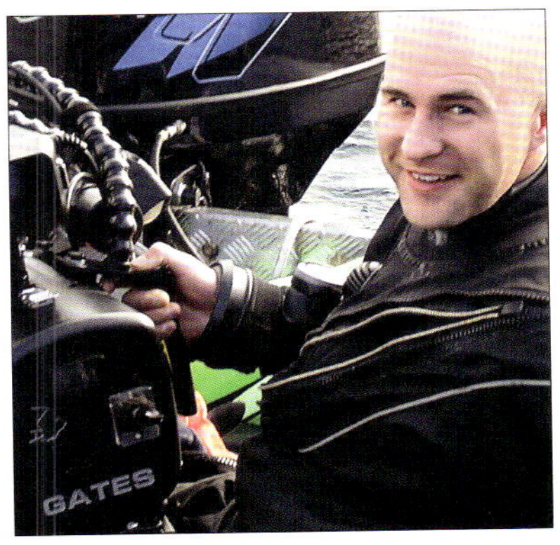

Shawn Harper is an adventurer who loves exploring our vast underwater world. Shawn started diving in 2001 with the American Academy of Underwater Sciences (AAUS) Scientific Diver Program at the University of Alaska Fairbanks (UAF). Since then he became a PADI/NAUI/AAUS diving instructor, dive medic, and crewmember at the Catalina Hyperbaric Chamber. He has also ventured into the technical diving realms of caves and re-breathers. Shawn participated in the Aleutians dives as part of his graduate studies at the UAF School of Fisheries and Ocean Sciences. While at UAF, he performed research dives in the Arctic, Antarctic, and Alaska, logging over 700 dives. Through these ventures he continues to learn about, document, and share his knowledge of the underwater world to promote quality education, outreach, research, and diving safety. Shawn's photos have appeared in Census of Marine Life materials, the *Alaska Coastal Calendar*, *Aurora* and *Alaska* magazines, *Anchorage Daily News*, *Fairbanks Daily News-Miner*, *The Antarctic Sun*, front cover of *Field Guide to Seaweeds of Alaska* and in *Surviving on the Foods and Water from Alaska's Southern Shores*, and exhibits at the Smithsonian Ocean Hall, Vancouver Museum, Museum of the Aleutians, and venues in Homer, Anchorage, and Fairbanks, Alaska.

Stephen Jewett

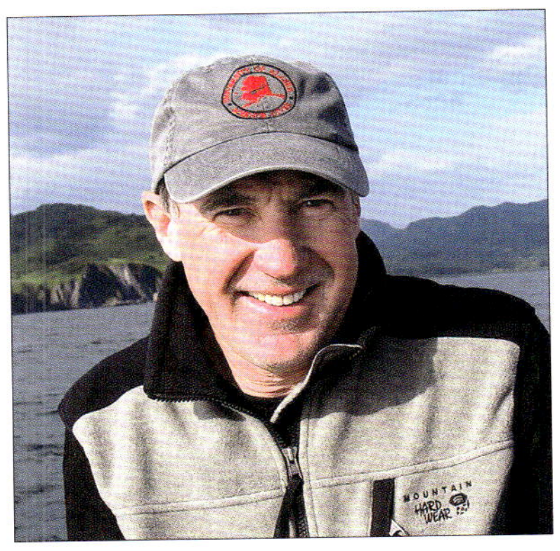

Stephen Jewett, Ph.D., has been a researcher at the University of Alaska Fairbanks (UAF) since 1974 and a research professor at the UAF Institute of Marine Science, School of Fisheries and Ocean Sciences, since 1998. Most of his research has focused on environmental impact assessment issues in nearshore marine waters, such as effects from commercial trawling, logging, dredging, mining, *Exxon Valdez* oil spill, and contamination from heavy metals and radionuclides. Stephen began diving in 1967 in Maine, where he grew up. Since then he has logged more than 2,000 dives, most in cold waters. Stephen has been the University of Alaska Dive Safety Officer since he founded the statewide Scientific Diving Program in 1988. Much of his research has utilized scientific diving techniques. Stephen has published 88 peer-reviewed journal articles and book chapters, and at this printing he is developing a color-illustrated field guide to Aleutian sea stars. His underwater photos have appeared in *Alaska* and *Aurora* magazines, on the front cover of *Field Guide to Common Marine Fishes and Invertebrates of Alaska*, in the books *The Bering Sea and Aleutian Islands: Region of Wonders* and *Ocean Treasure: Commercial Fishing in Alaska*, in the *Alaska Coastal Calendar* and *Visions of Undersea Alaska Engagement Calendar*, and on the cover of an Alaska Sea Grant biennial report.

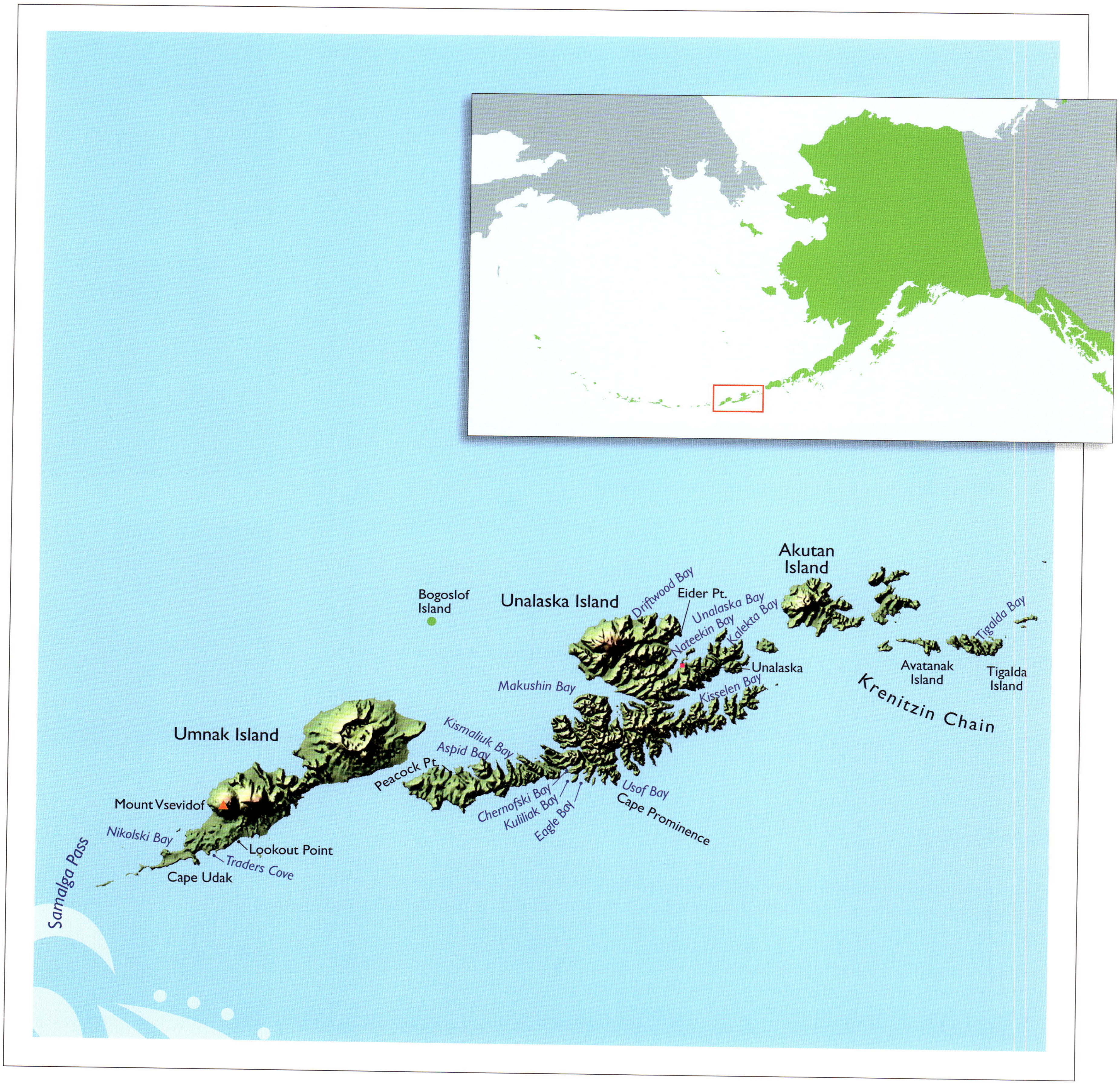

Akutan Island
Bogoslof Island
Unalaska Island
Driftwood Bay
Eider Pt.
Unalaska Bay
Nateekin Bay
Kalekta Bay
Tigalda Bay
Unalaska
Avatanak Island
Tigalda Island
Krenitzin Chain
Makushin Bay
Kisselen Bay
Umnak Island
Kismaliuk Bay
Aspid Bay
Peacock Pt.
Mount Vsevidof
Usof Bay
Cape Prominence
Chernofski Bay
Kuliliak Bay
Eagle Bay
Nikolski Bay
Lookout Point
Traders Cove
Cape Udak
Samalga Pass

Fox Islands

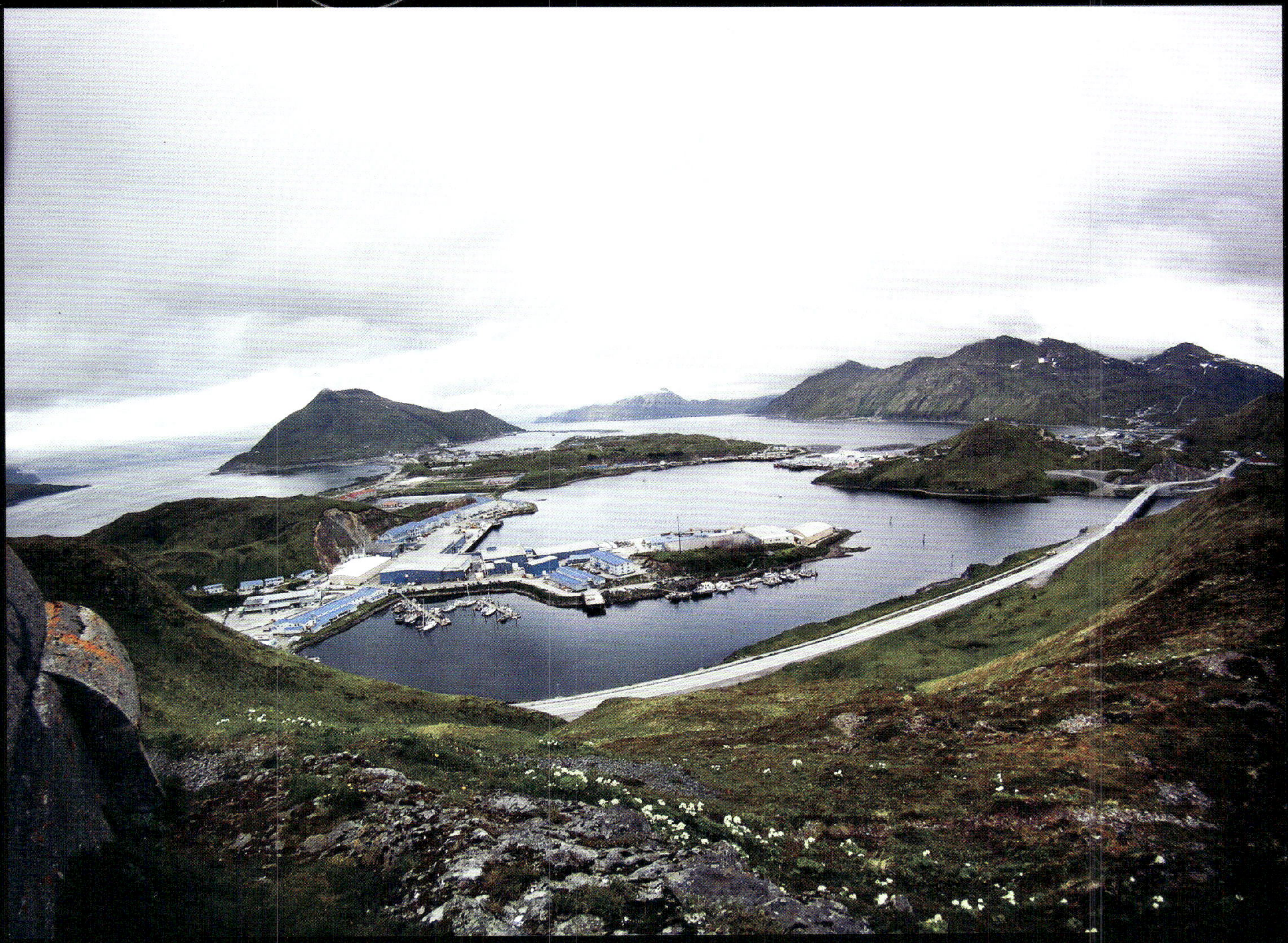

The community of Unalaska is approximately 800 miles southwest of Anchorage, Alaska. The Unangan people were the first to inhabit Unalaska Island, which they named "Ounalashka" or "near the peninsula." Artifacts dating back roughly 9,000 years suggest that Unalaska is one of the oldest continually inhabited islands in the United States. The Russian influence is best viewed by touring the Holy Ascension Russian Orthodox Cathedral, one of the oldest cruciform-style Russian churches in the country. *R. Brewer photo.*

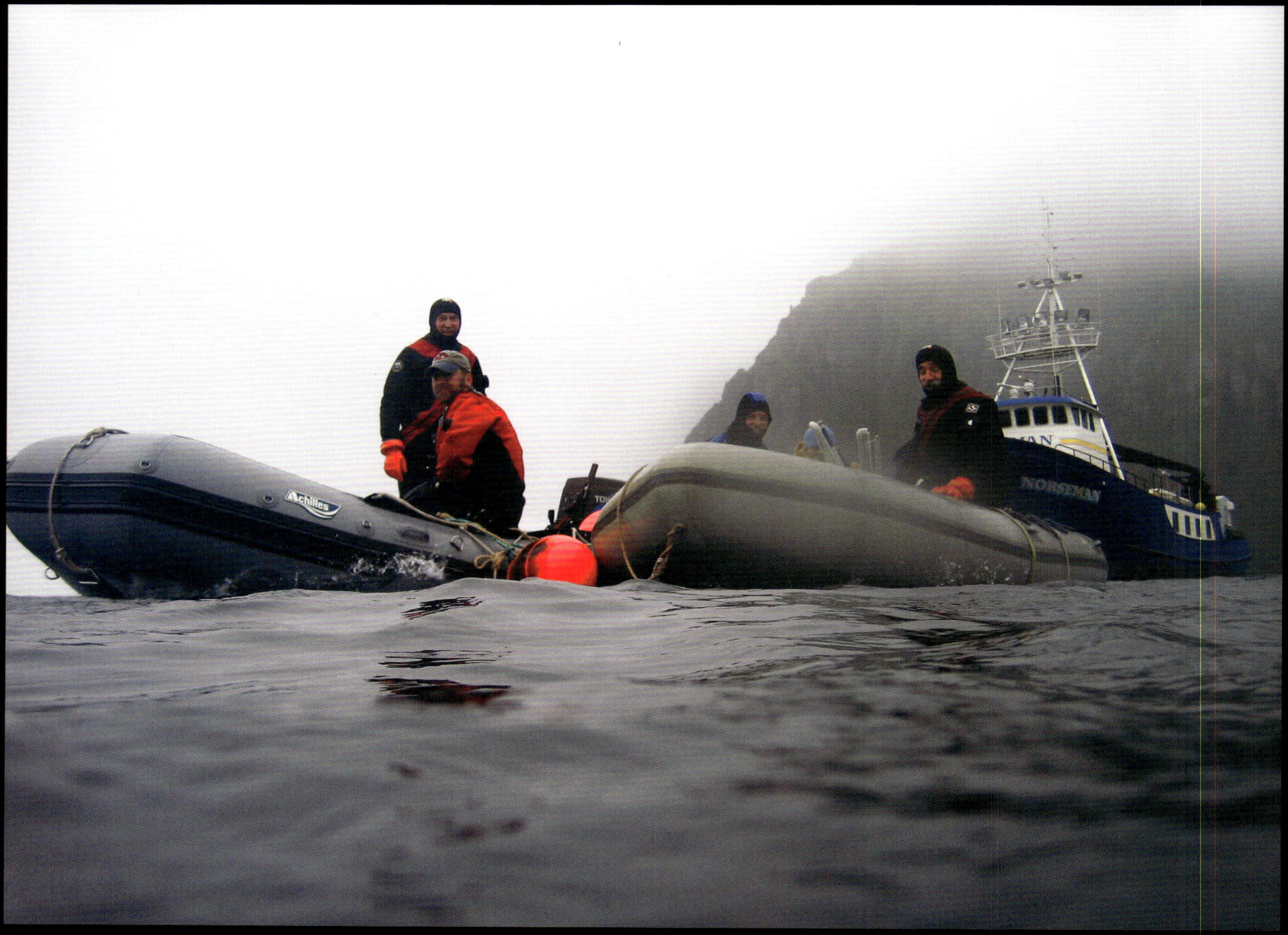

Divers prepare for sampling near the community of Unalaska. They transited the Aleutians on the 108 ft R/V *Norseman* (background), and performed all dive operations off inflatable skiffs. Although the weather can be very mild in summer (45-57°F, 7-14°C), fog occurs on average 151 days a year, and is prominent between June and September. *Kalekta Bay, Unalaska Island, H. Chenelot photo.*

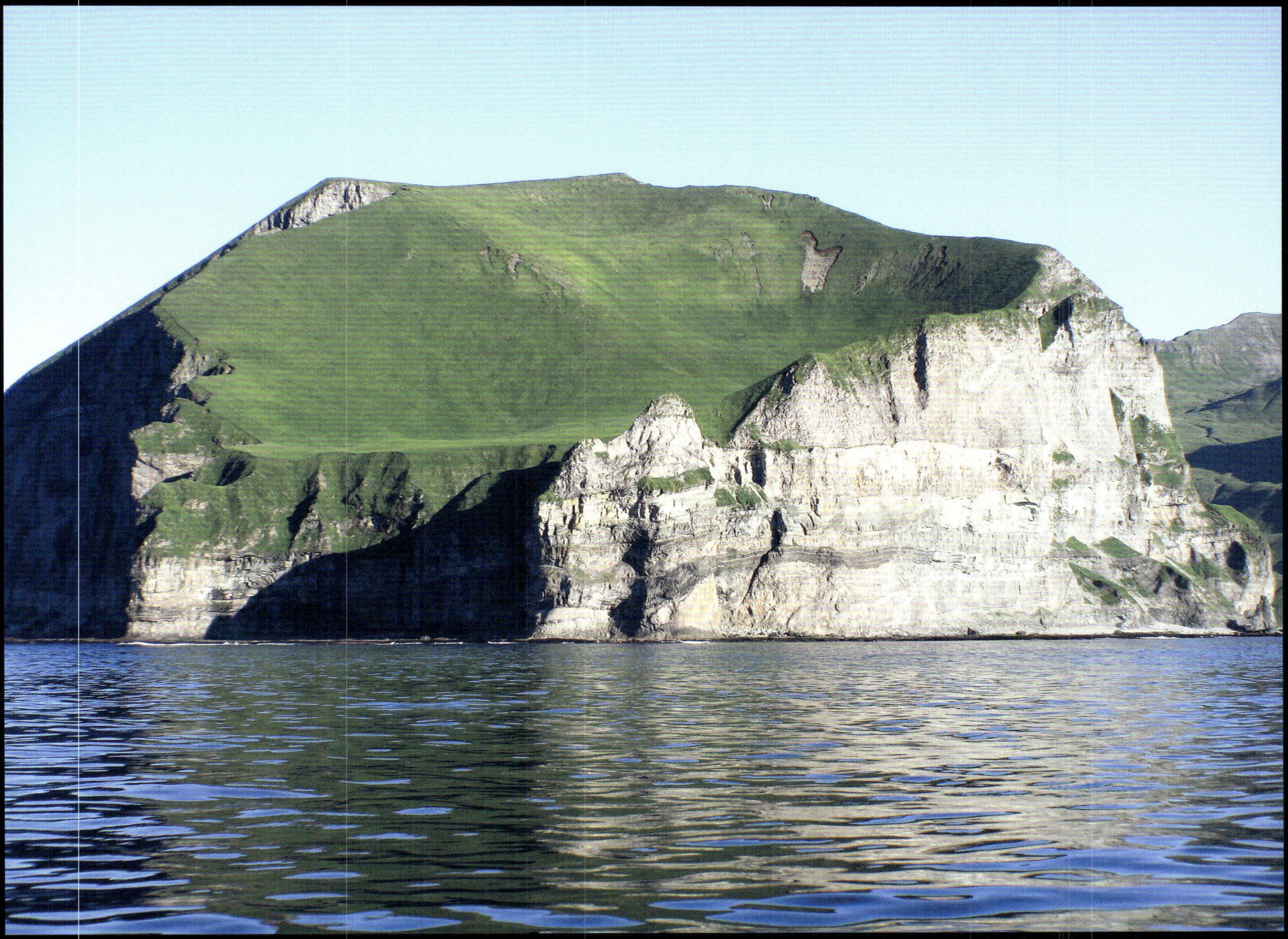

The Aleutian landscape is a mix of geological and biological wonders. The rock layers are the result of multiple volcanic eruptions over 10,000 years. The coasts are rugged and dangerous, often rising abruptly to steep mountains. Most islands are devoid of trees, but are instead blanketed by a luxuriant, dense growth of shrubs, grasses, and other flowering plants. *Cape Prominence, Unalaska Island, M. Lindeberg photo.*

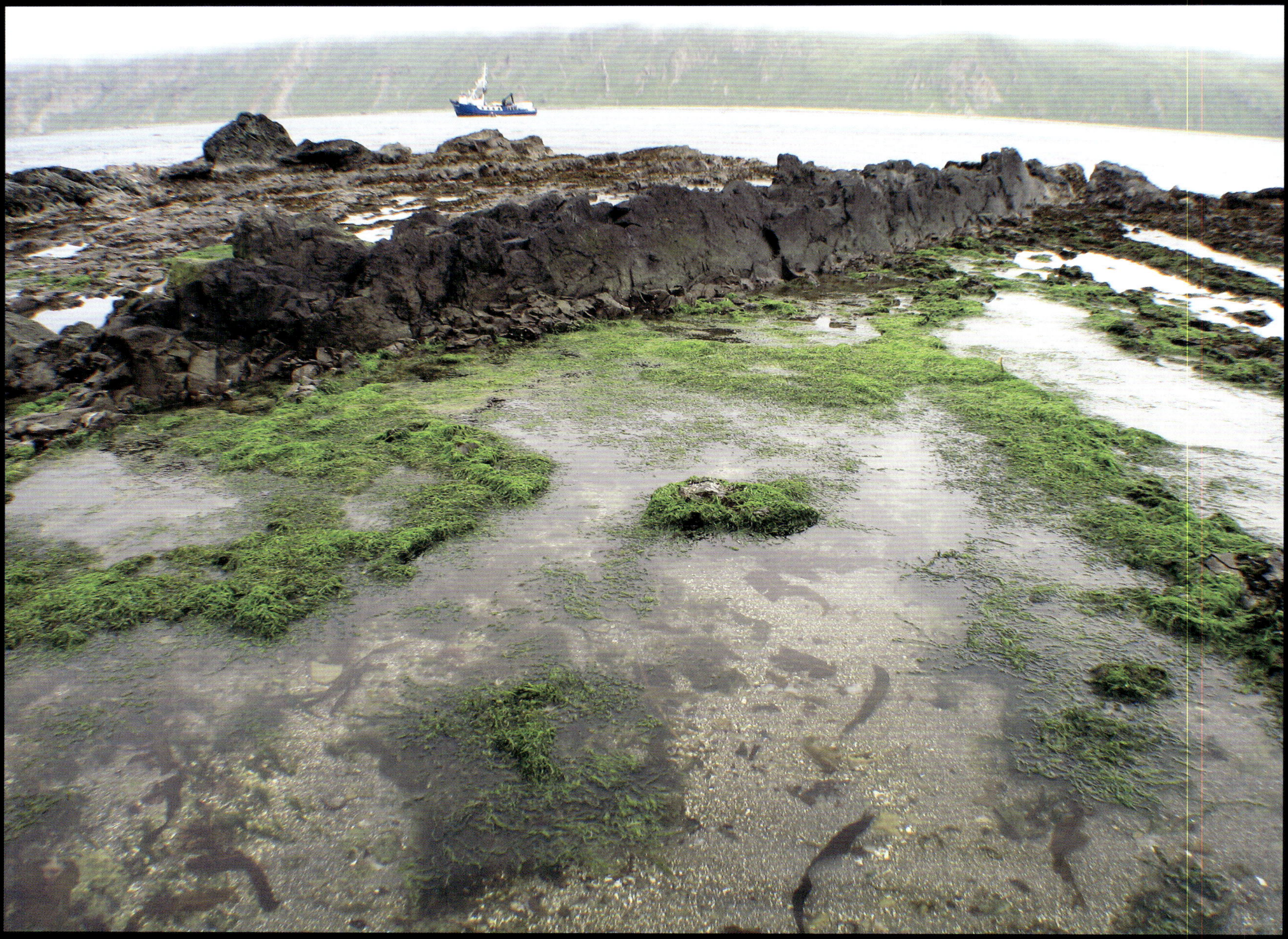

Algae are diverse in the intertidal zone. The linear rock projections, called dikes, are the result of magma intrusion into fractures in the bedrock. The intrusions eroded more slowly than the surrounding bedrock. *Aspid Bay, Unalaska Island, M. Lindeberg photo.*

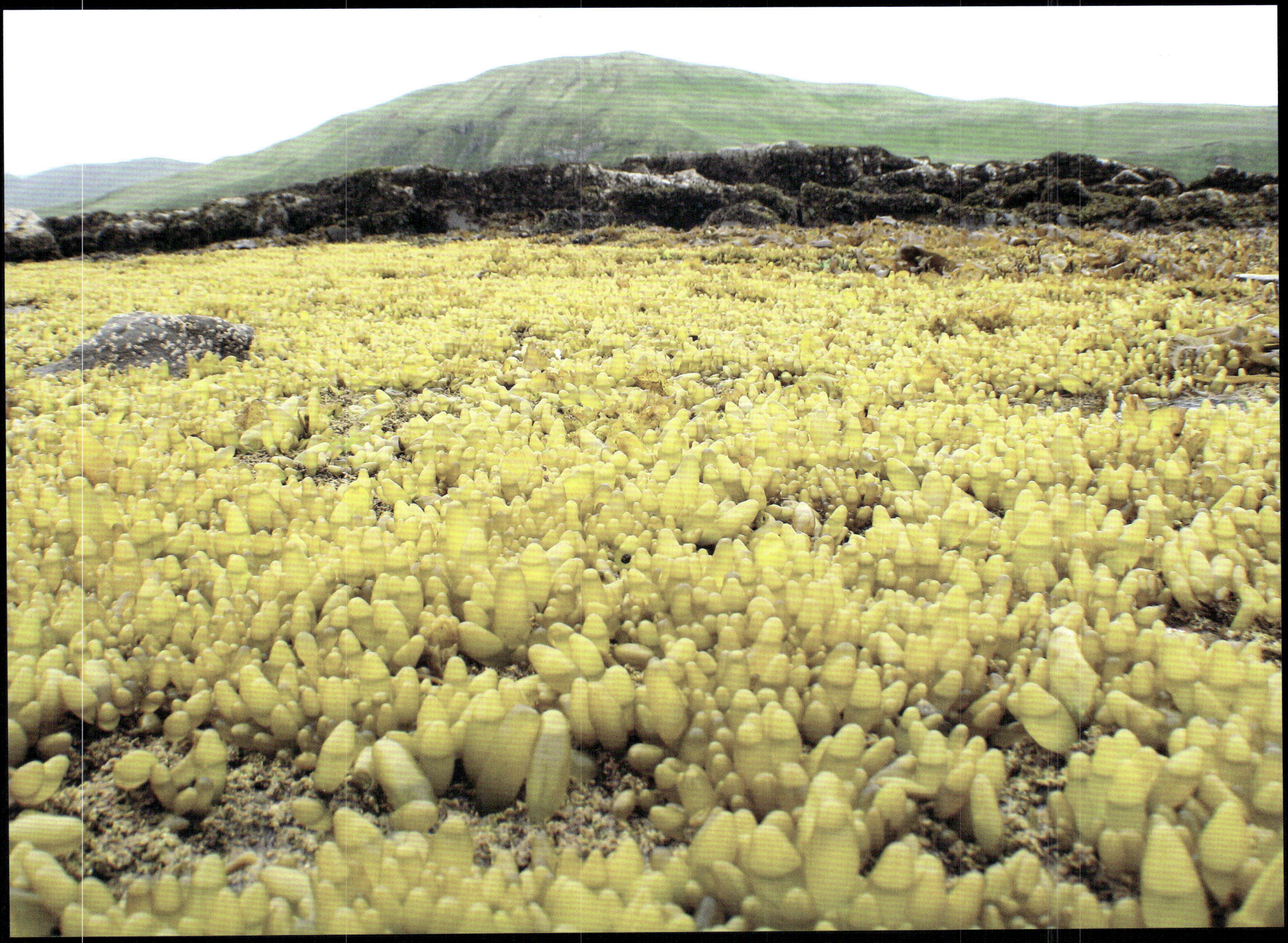

The sea sac (*Halosaccion glandiforme*) is a red seaweed found in shallow, sloping intertidal areas. This species is distinguished by the hollow sacs, 2-6 in (5-15 cm) tall, filled with seawater with an air bubble at the tip. The sea sac ranges in color from yellowish-brown in direct sunlight to reddish-purple in shade. In the Aleutians it commonly forms expansive carpets of short, yellow individuals. *Kismaliuk Bay, Unalaska Island, M. Lindeberg photo.*

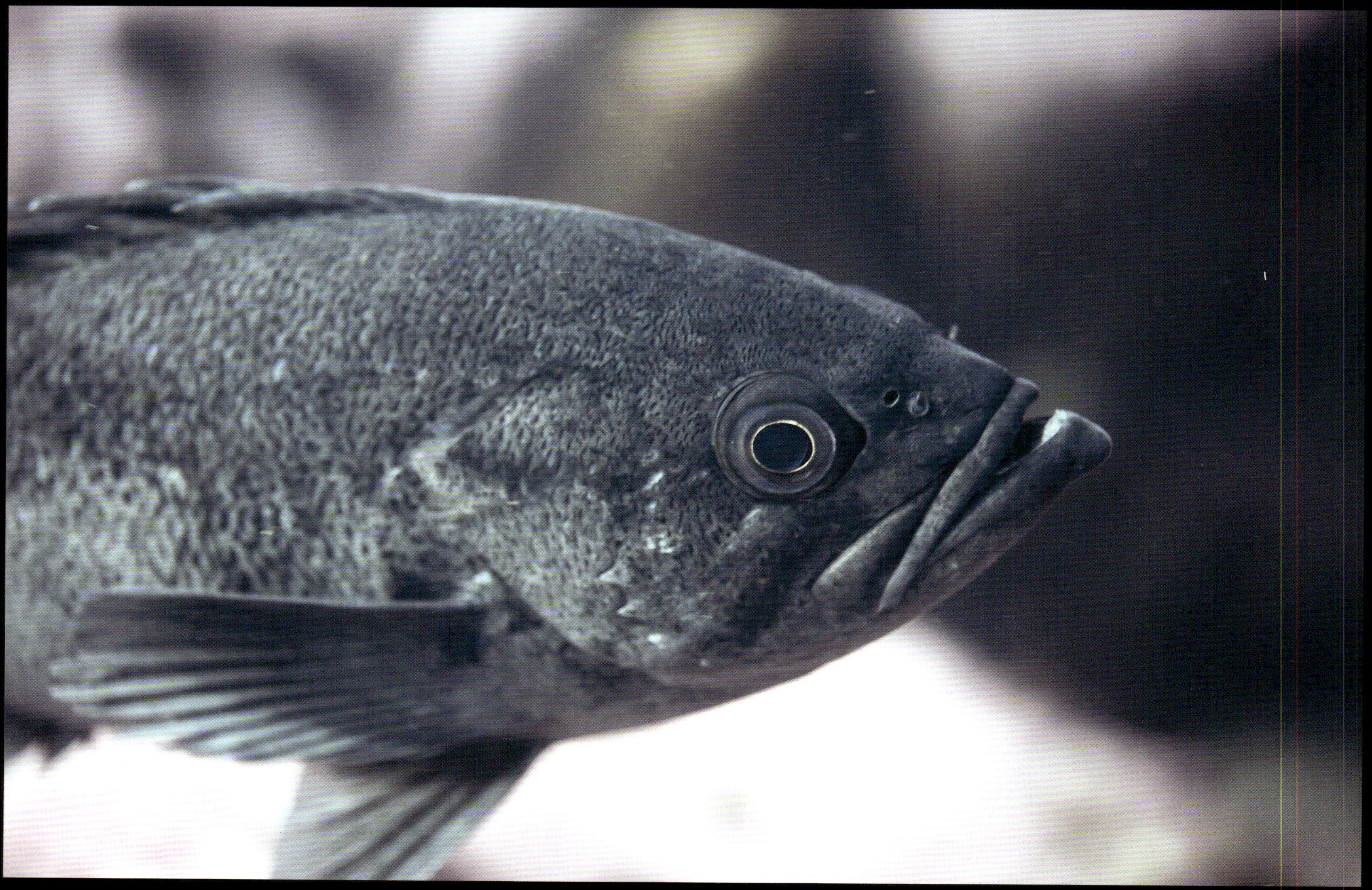

Dusky rockfish (*Sebastes ciliatus*) are ubiquitous throughout the northern seas. There are 32 species of rockfishes in Alaska, most of which are important to commercial fishing and sportfishing industries. Some species of rockfish reach over 100 years in age. *Eider Point, Unalaska Island, R. Brewer photo.*

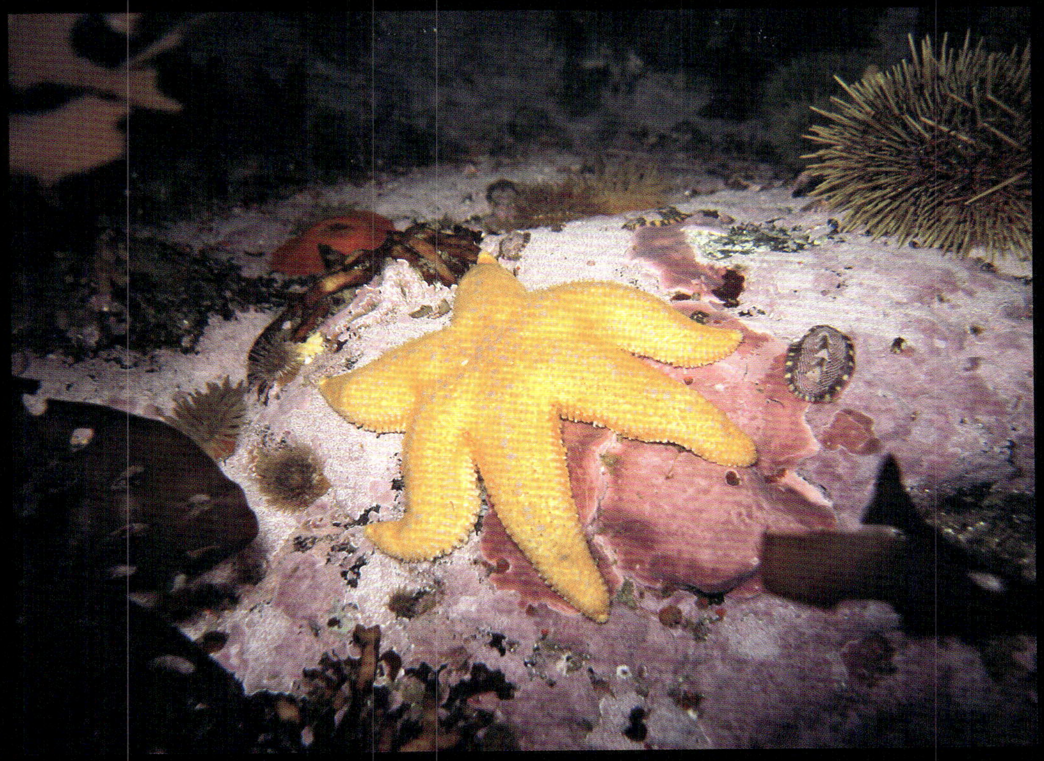

The Aleutian knobby six-rayed star (*Leptasterias dispar*) can be found in all colors throughout the Aleutians. The aboral surface (back or top) of this sea star is covered with small spines and pedicellariae (little pinchers). In many species of six-armed sea stars, the female broods her young underneath her arched body until the juveniles crawl away to fend for themselves. *Nateekin Bay, Unalaska Island, R. Brewer photo.*

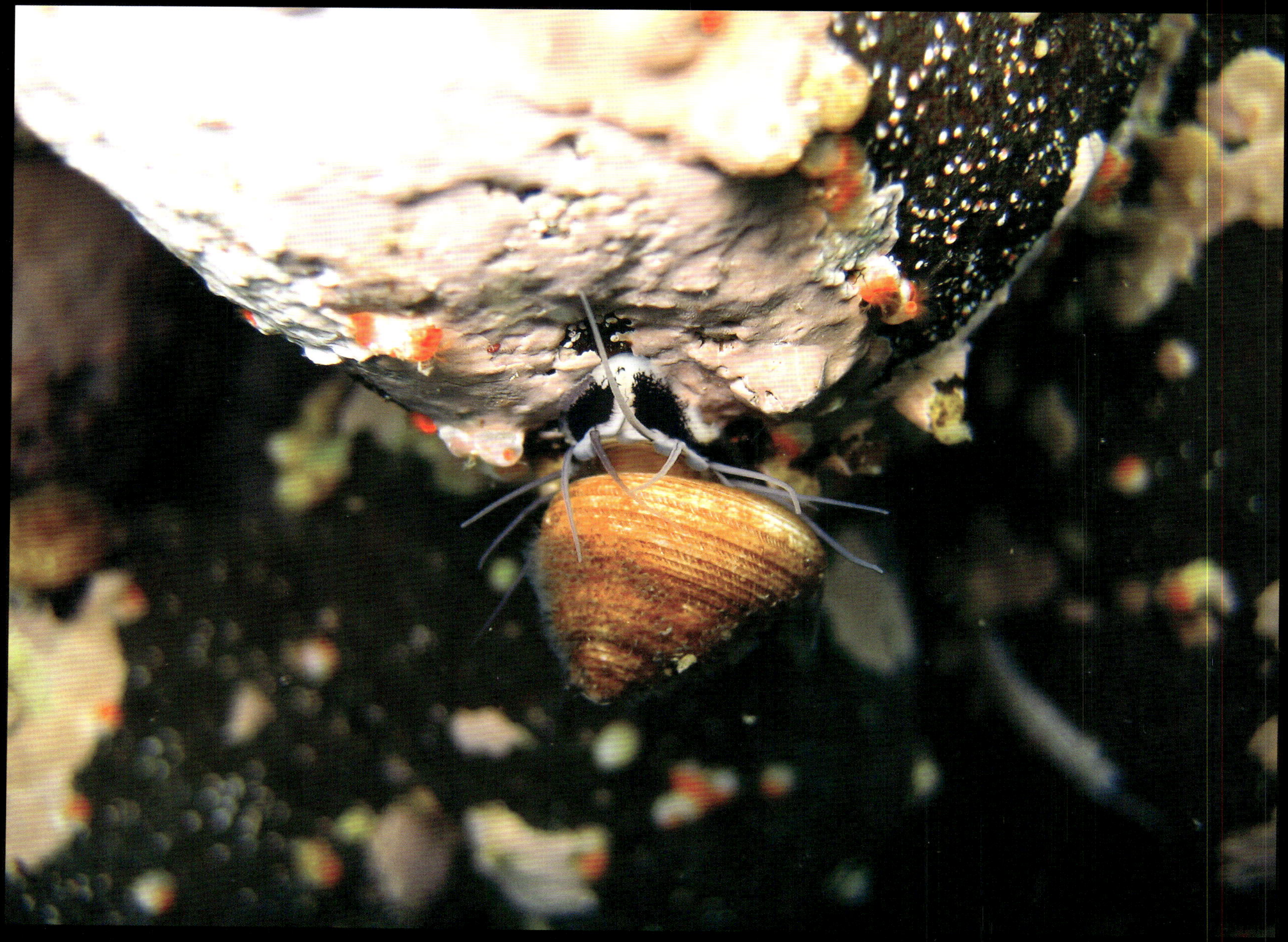

Puppet margarite snails (*Margarites pupillus*) are tiny, active organisms found on rocks and kelp. Note the delicate fleshy filaments that serve as sense organs. Puppet margarites feed on hydroids, bryozoans, diatoms, detritus, and kelp. This species occurs throughout the Aleutian Islands to southern California, from the intertidal zone to a depth of 1,000 ft (300 m). *Eider Point, Unalaska Island, H. Chenelot photo.*

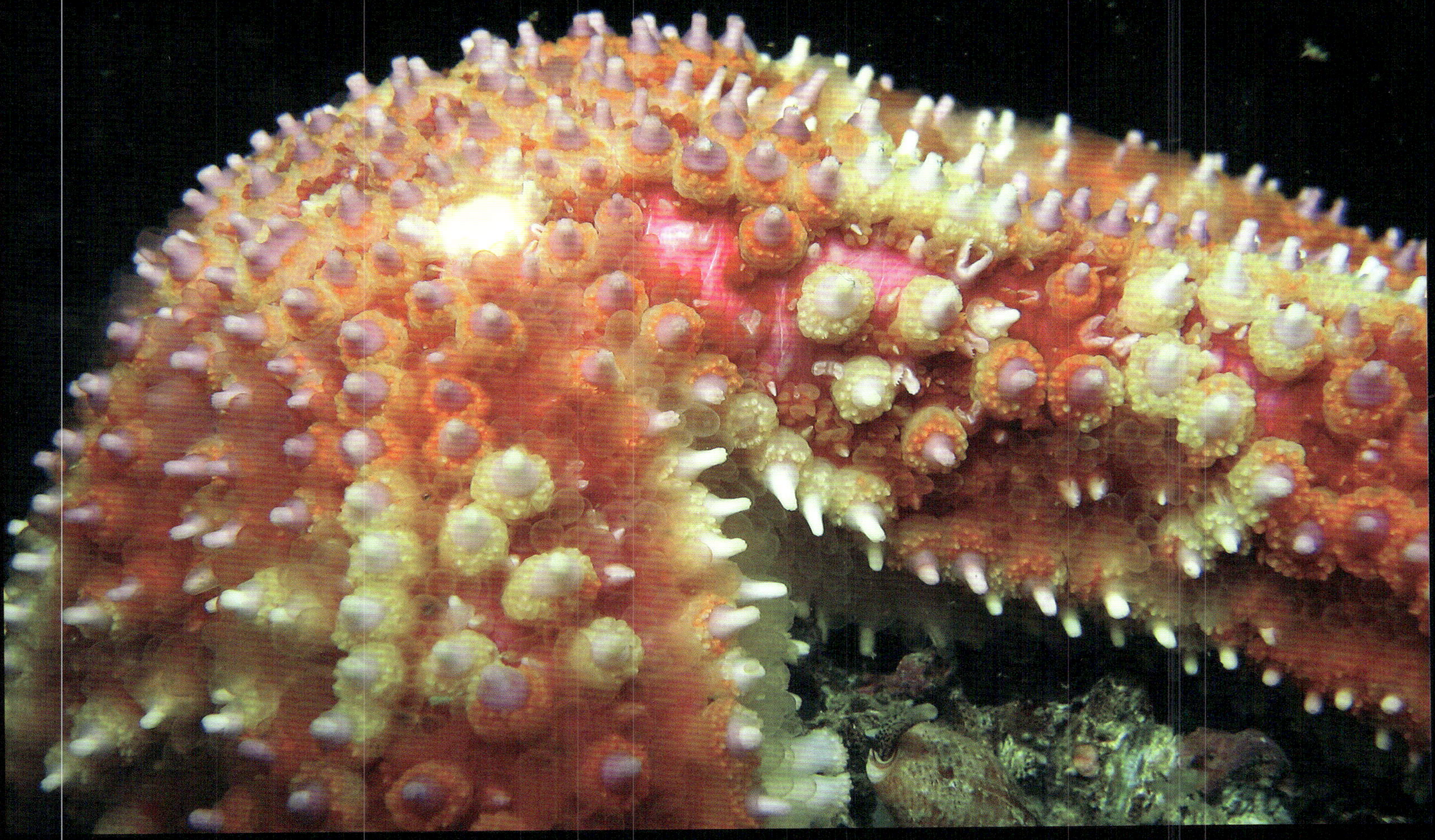

The rainbow star (*Orthasterias koehleri*) has banded red and white rays (arms), with prominent white spines and a conspicuous ruff of pedicellariae or small pinchers. Rainbow stars eat bivalves, chitons, and sea squirts and are taken as prey by sea otters (*Enhydra lutris*). This slender-armed sea star is found from Alaska to northern Mexico. *Eagle Bay, Unalaska Island, H. Chenelot photo.*

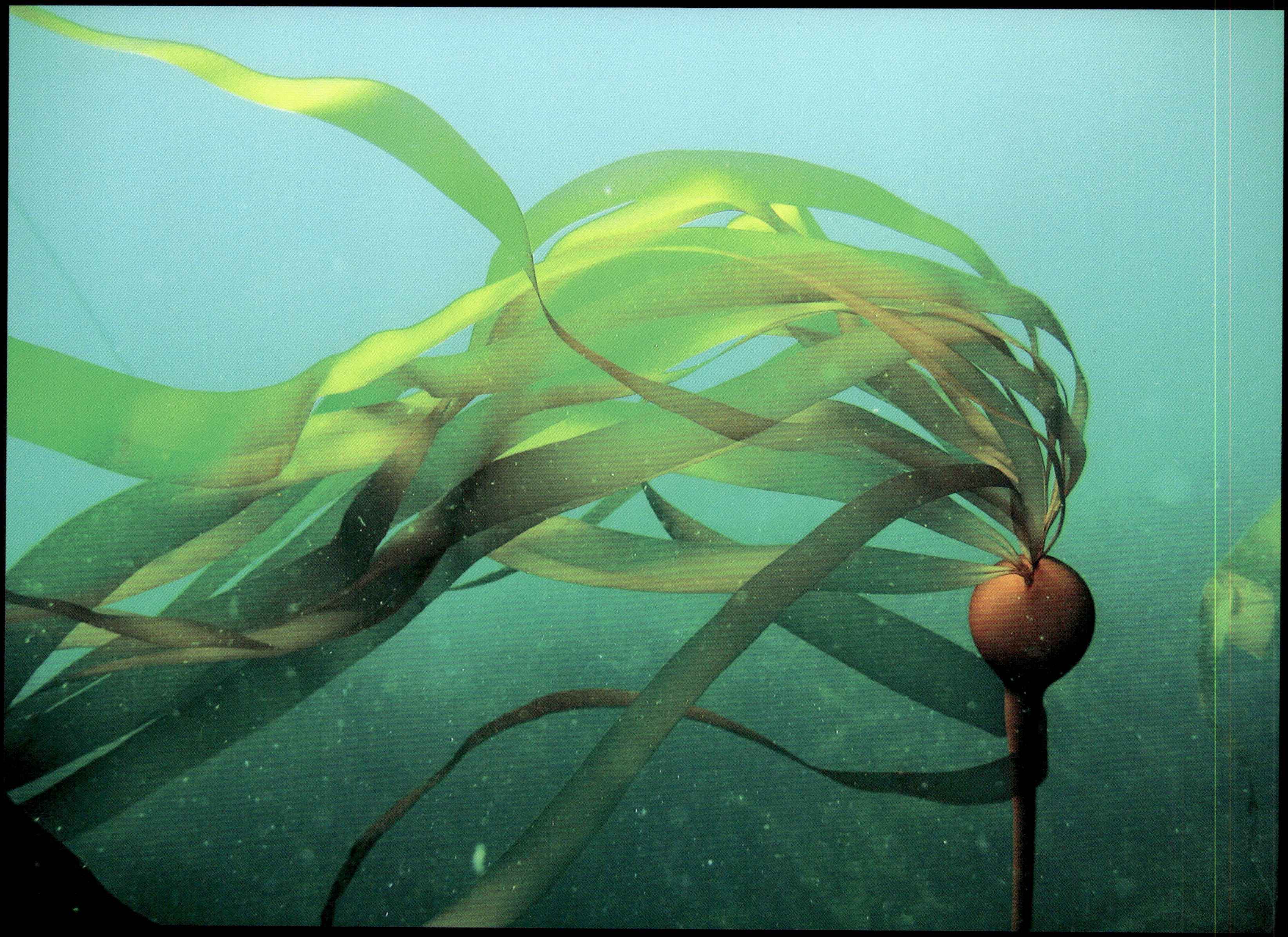

A solitary bull kelp (*Nereocystis luetkeana*) sways in a mild current. This annual seaweed is fast growing (up to 9.8 in per day, 25 cm) and can reach 115 ft (35 m) in height. It plays an important role as habitat for other organisms. The large bulbous pneumatocyst (float) contains carbon monoxide. Alaska Natives sometimes use bull kelp floats to hold fresh water or fish oil. It occurs from the eastern Aleutians to the central California coast. *Kuliliak Bay, Unalaska Island, H. Chenelot photo.*

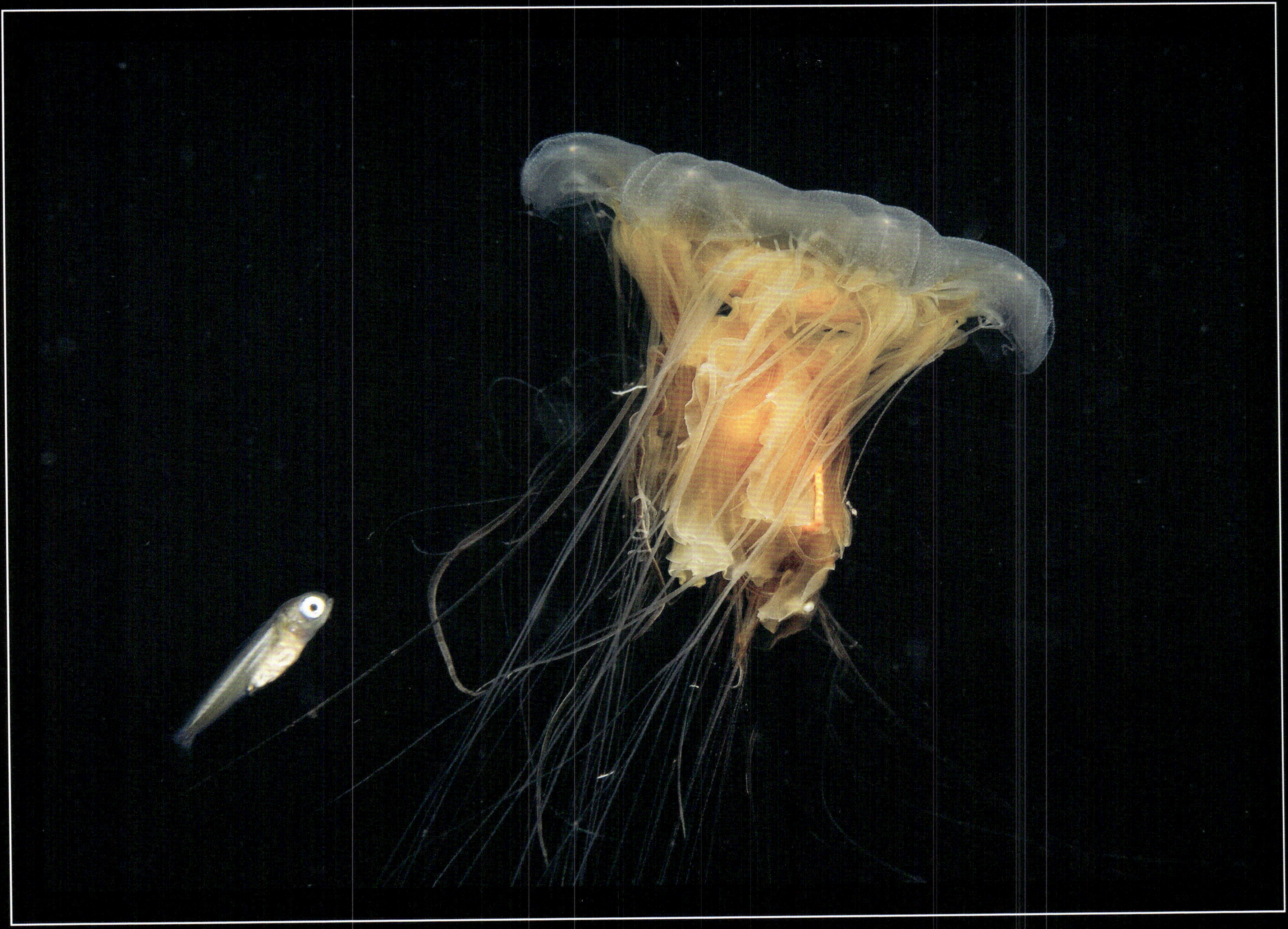

The lion's mane jellyfish (*Cyanea capillata*) is one of the largest jellyfish in the world despite its short life of less than a year. Its brownish-red bell can measure up to 6.5 ft (2 m) in diameter, and several hundred tentacles reach lengths of 30 ft (9 m). This jellyfish has potent stinging cells on its tentacles and divers avoid it whenever possible, though the tentacles may shelter nimble fish species. It ranges from Alaska to Oregon. *Usof Bay, Unalaska Island, S. Harper photo.*

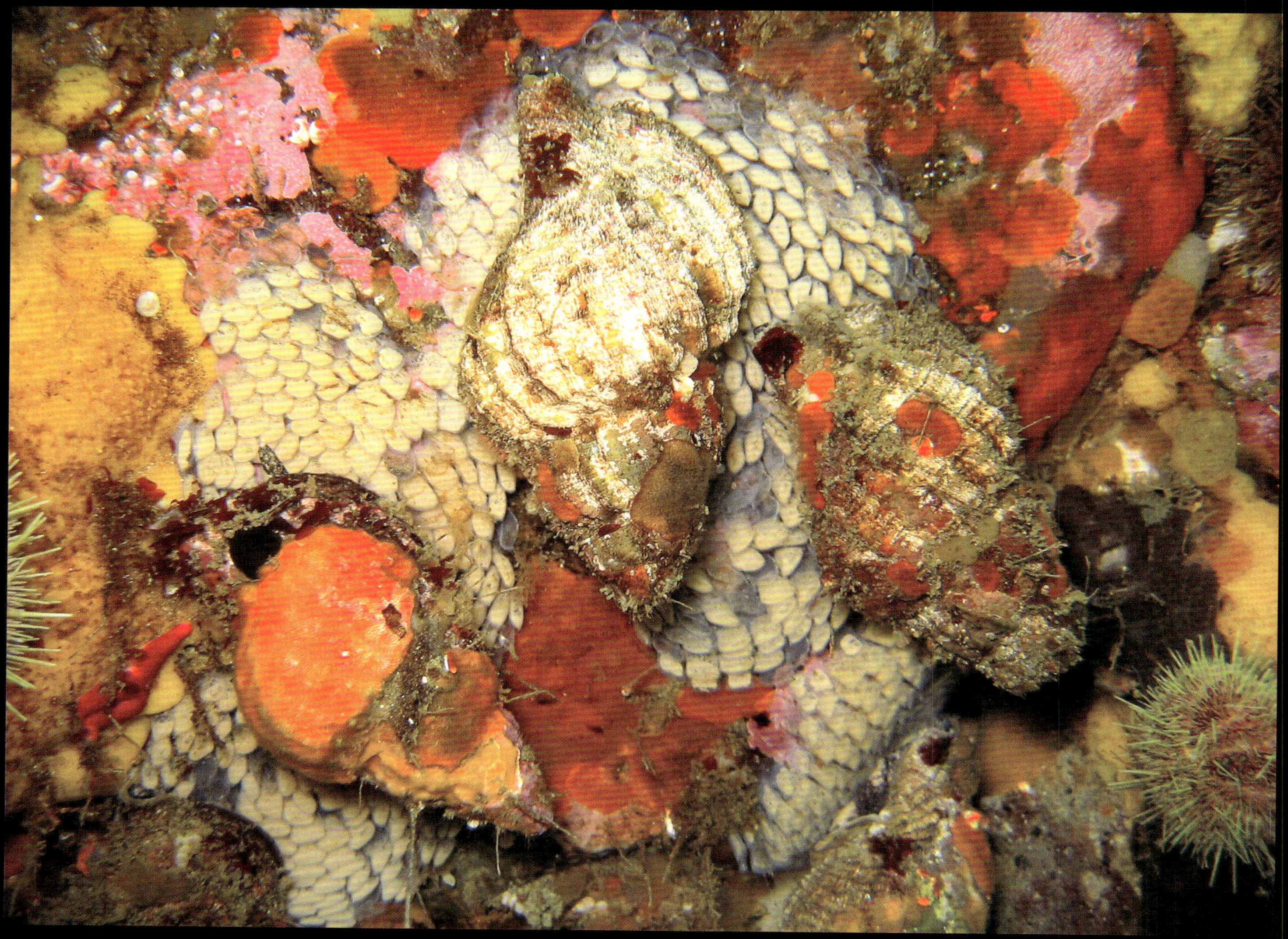

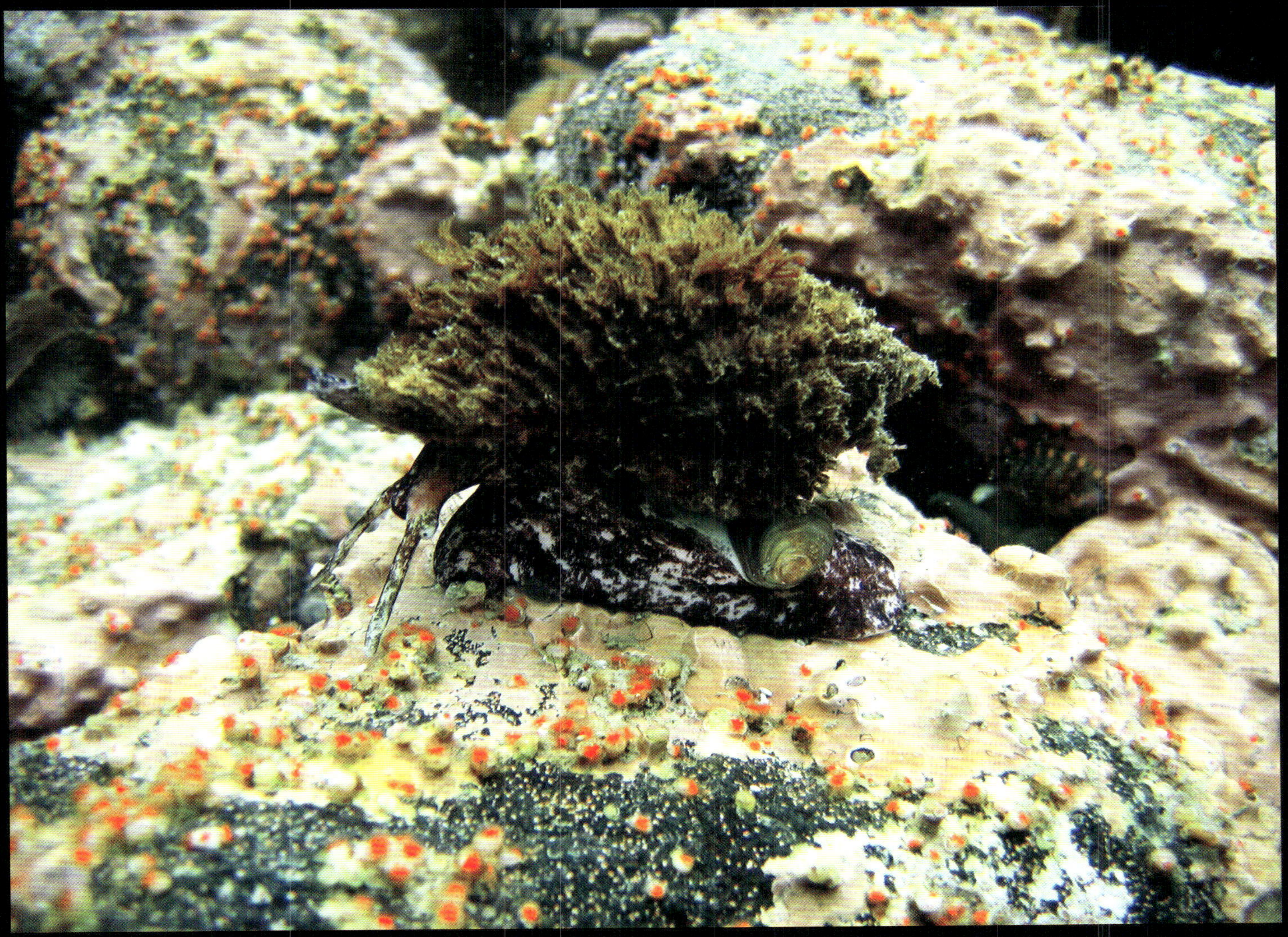

With a large brown shell covered with hairs, the Oregon or hairy triton (*Fusitriton oregonensis*) is a conspicuous inhabitant of the North Pacific nearshore and is often seen tending spiral-shaped egg masses. *Fusitriton oregonensis* is a carnivore and scavenger, feeding on a myriad of invertebrates. *Left: three tritons, Amchitka Island, H. Chenelot photo. Right: single triton, Eider Point, Unalaska Island, H. Chenelot photo.*

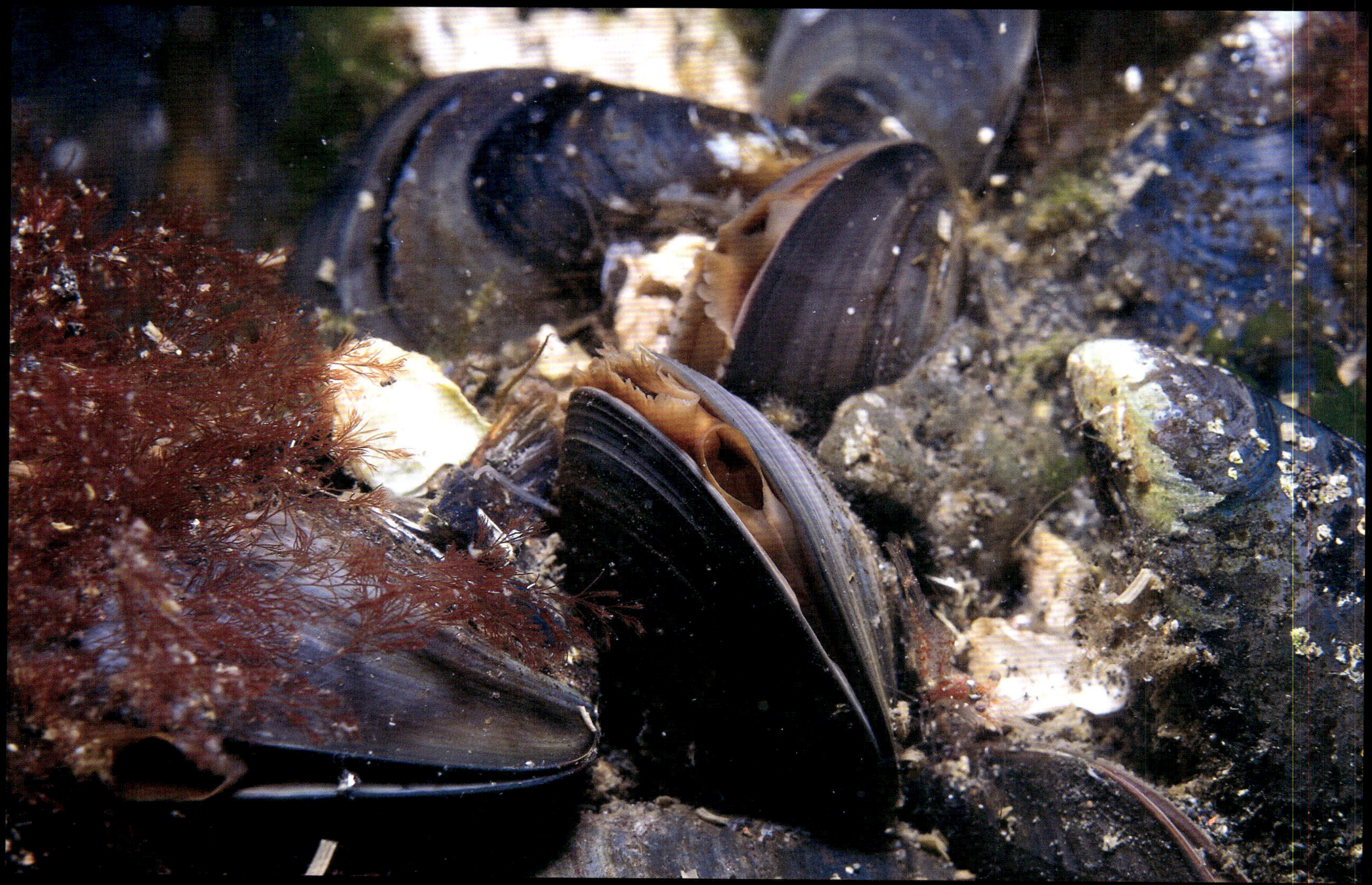

The blue mussel (*Mytilus trossulus*) has a blue-black shell and secretes strong, silken byssal threads to attach to the substrate or to other mussels. Blue mussels are an important food source in many rural Alaska communities. In rare cases, paralytic shellfish poisoning (PSP) makes them unsafe for consumption. *Kismaliuk Bay, Unalaska Island, S. Harper photo.*

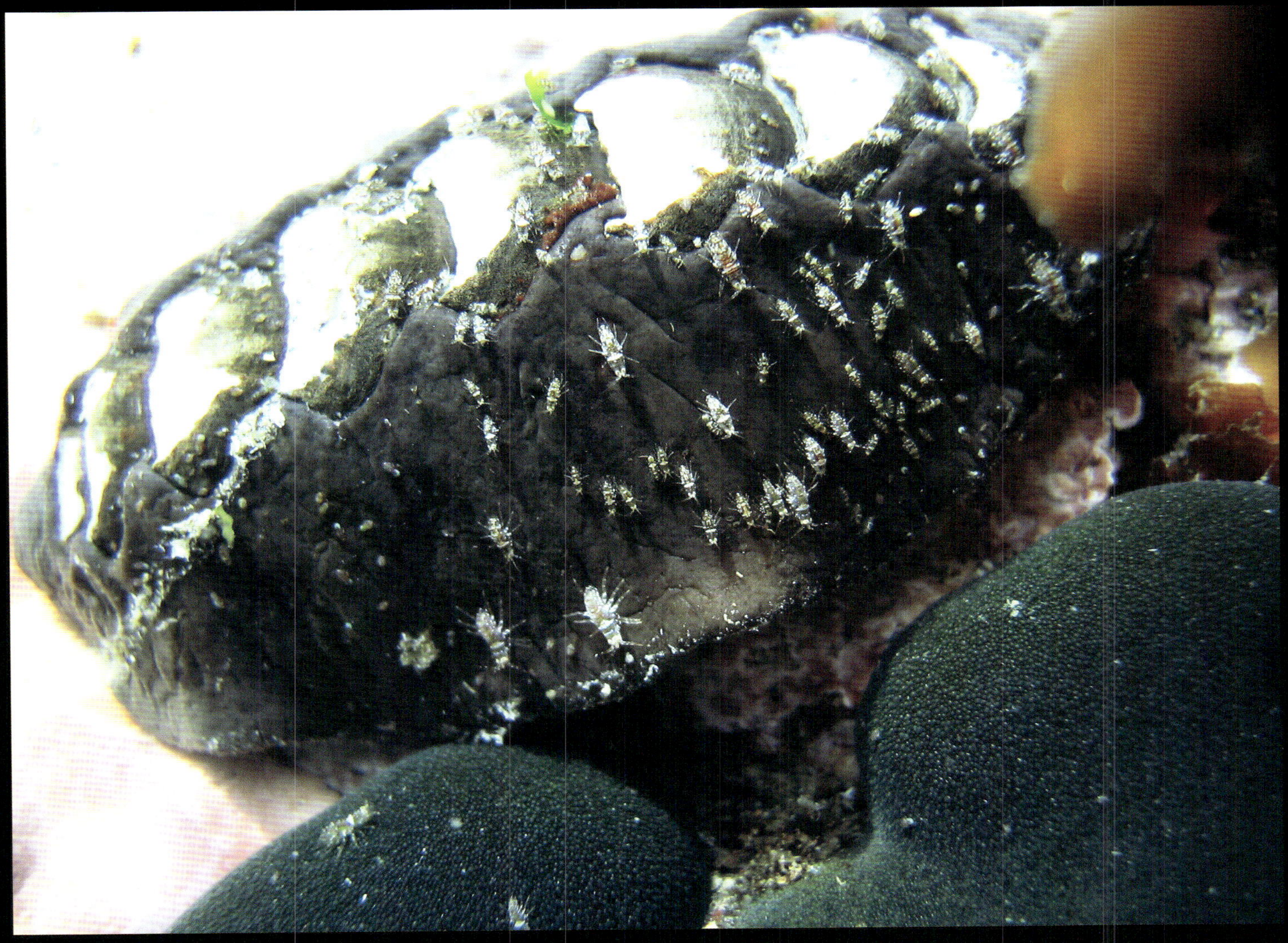

The black leather chiton, or black Katy (*Katharina tunicata*), is usually found in the shallow subtidal and intertidal zone from Russia and northern Alaska to southern California. The girdle (mantle of muscular tissue surrounding the plates) has a distinct leathery texture and covers most of the white plates. The black Katy is a subsistence food of Native Alaskans. Aleuts refer to it as "bidarka" because it resembles an overturned bidarka or kayak. Archaeological research shows that *K. tunicata* was an important food for early Alaskans as well. The chiton in the photo hosts many amphipods. *Unalaska Island, S. Jewett photo.*

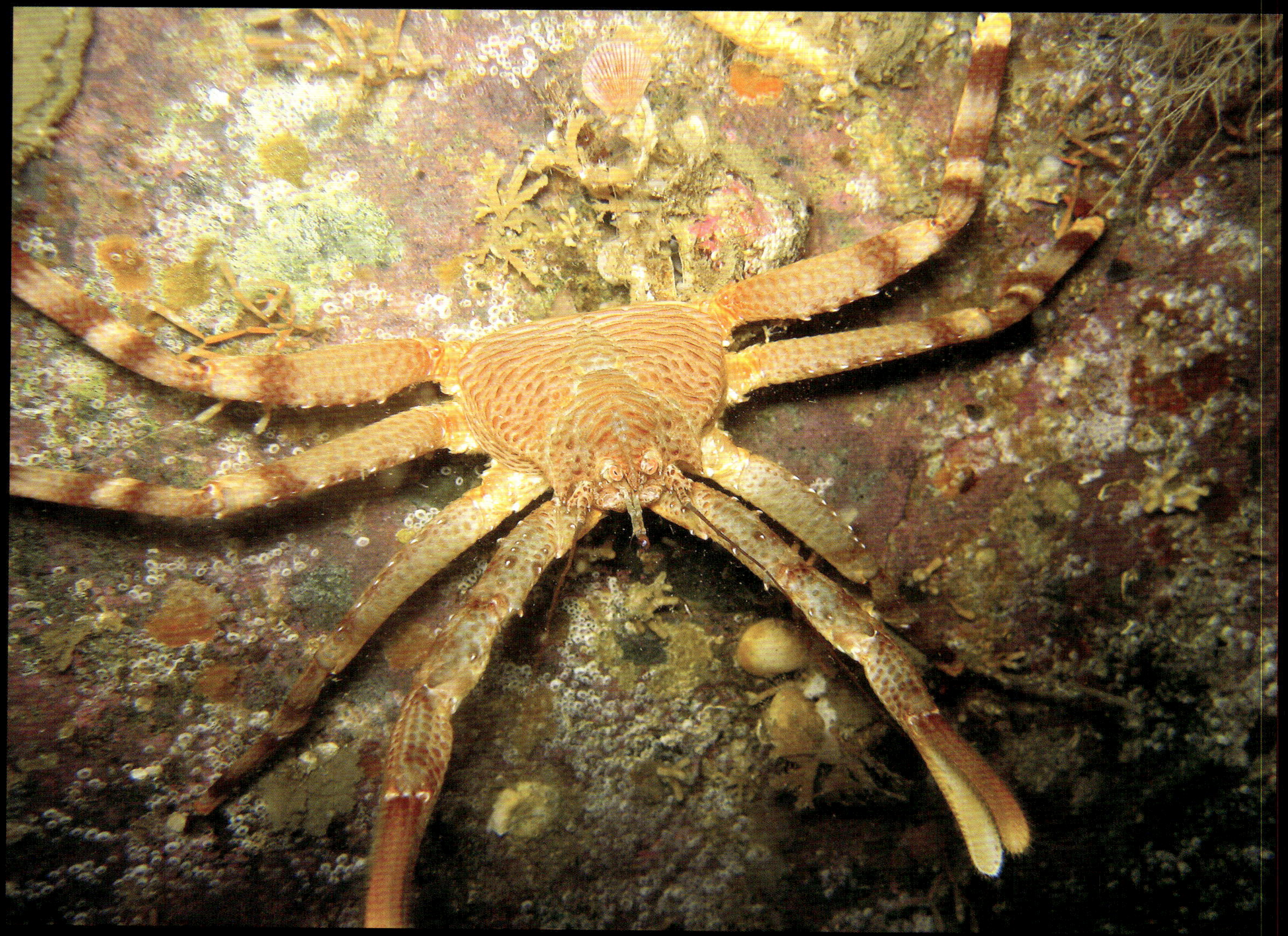

The scaled crab (*Placetron wosnessenskii*) is named for the scale-like ridges on its legs and carapace. Scaled crabs are adept at blending in with rocky habitats, but if discovered they will move quickly to find a new hiding spot. This species has spatulate pincers, instead of the more common sharp-edged chelae (claws). It is distributed throughout the Aleutians to Puget Sound, in shallow waters to 360 ft (110 m). *Kismaliuk Bay, Unalaska Island, R. Brewer photo.*

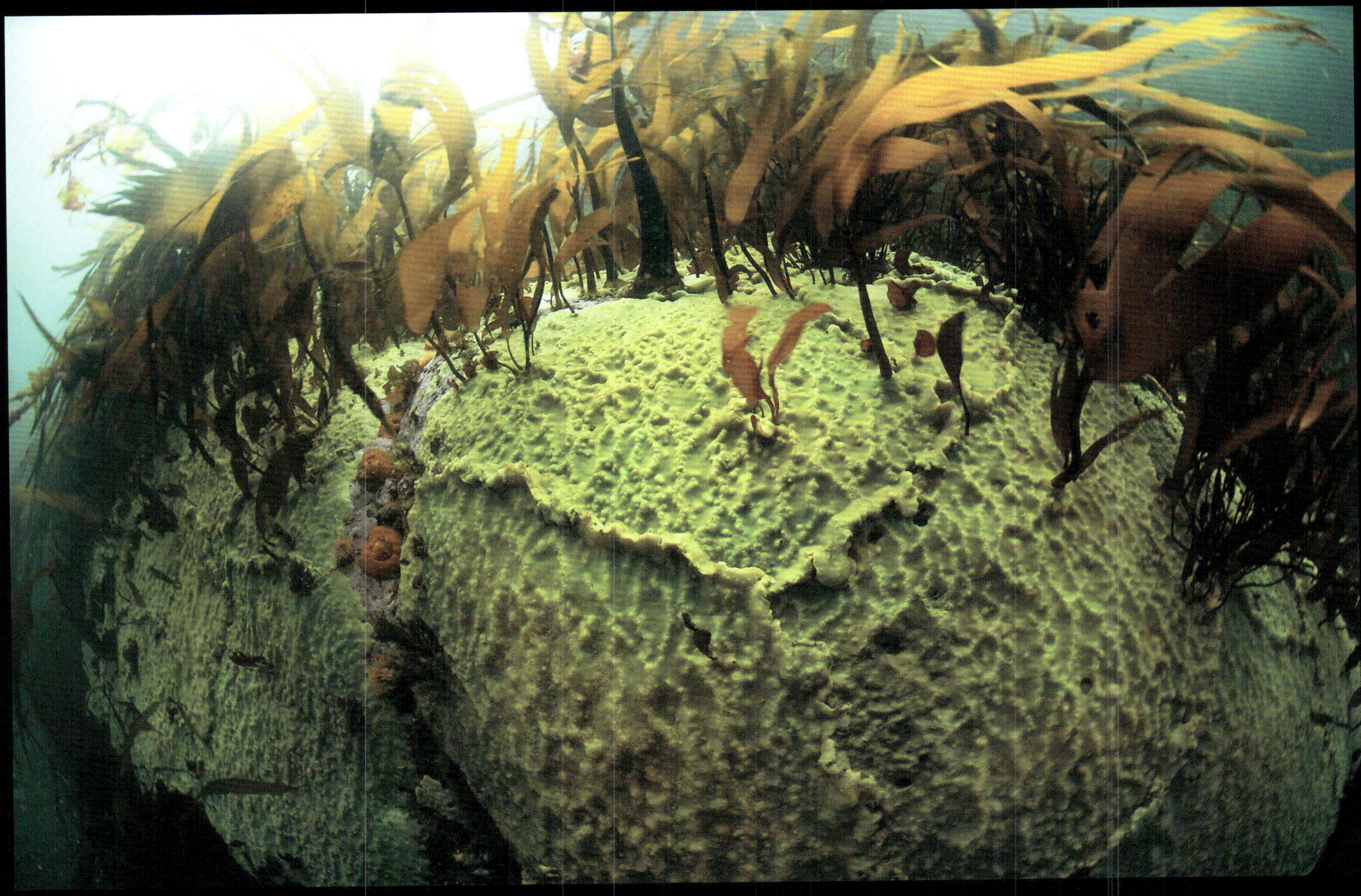

The breadcrumb sponge (*Halichondria panicea*) is one of the few sponge species in Alaska that is readily seen by beachcombers and divers alike. This encrusting species can be found in large patches and is eaten mainly by sea stars and nudibranchs. In this photo, breadcrumb sponge surrounds juvenile laminarian kelps. *Nateekin Bay, Unalaska Island, S. Harper photo.*

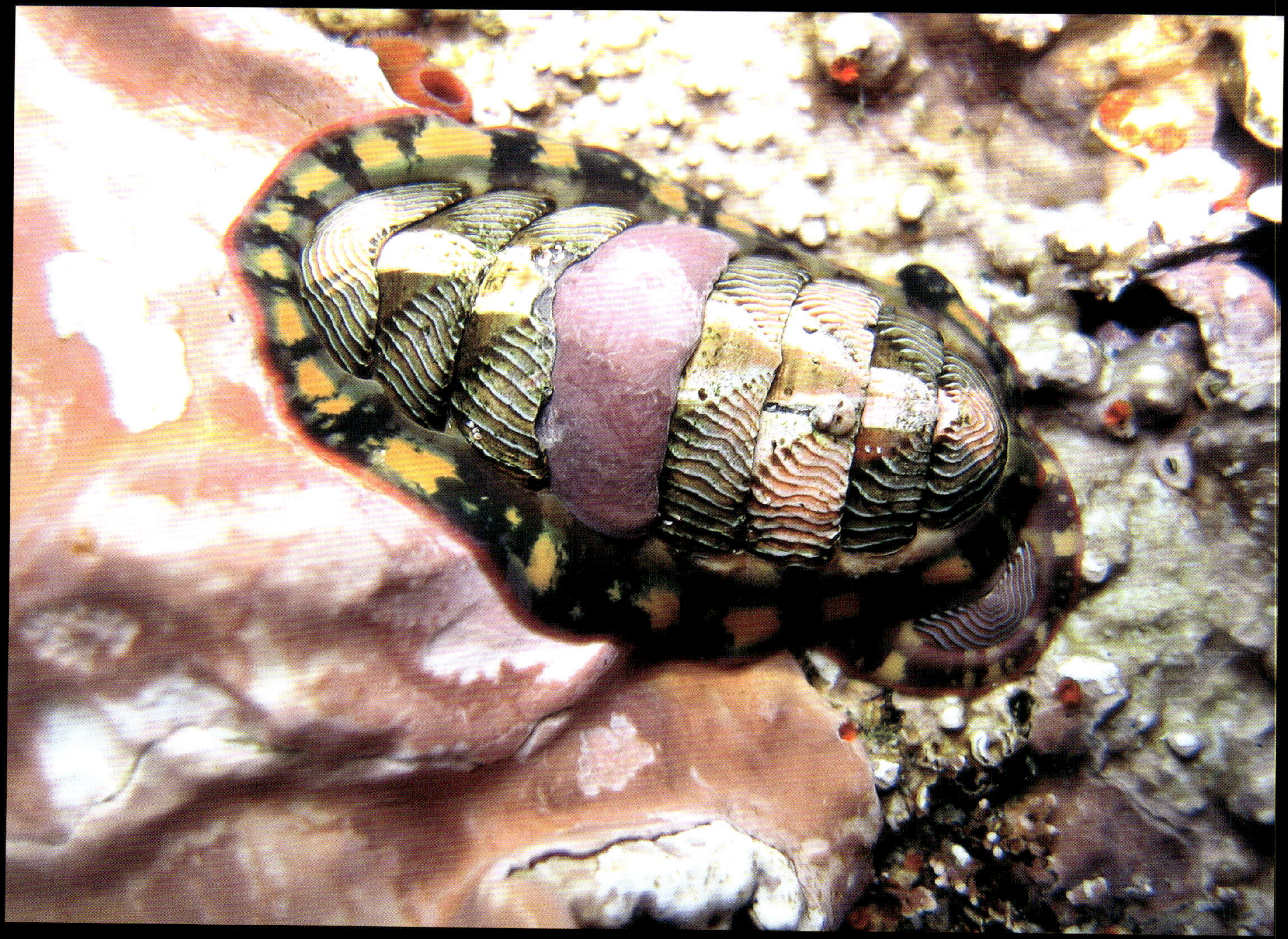

The lined chiton (*Tonicella lineata*) is distinguished by the wavy lines at the edges of the valves (shells). This chiton has many color variations, which are linked to the diet of each individual. Most chitons are herbivores and forage on algae (including calcified coralline algae) that cover rocks. The lined chiton is found from California to Alaska to Japan. *Driftwood Bay, Unalaska Island, S. Jewett photo.*

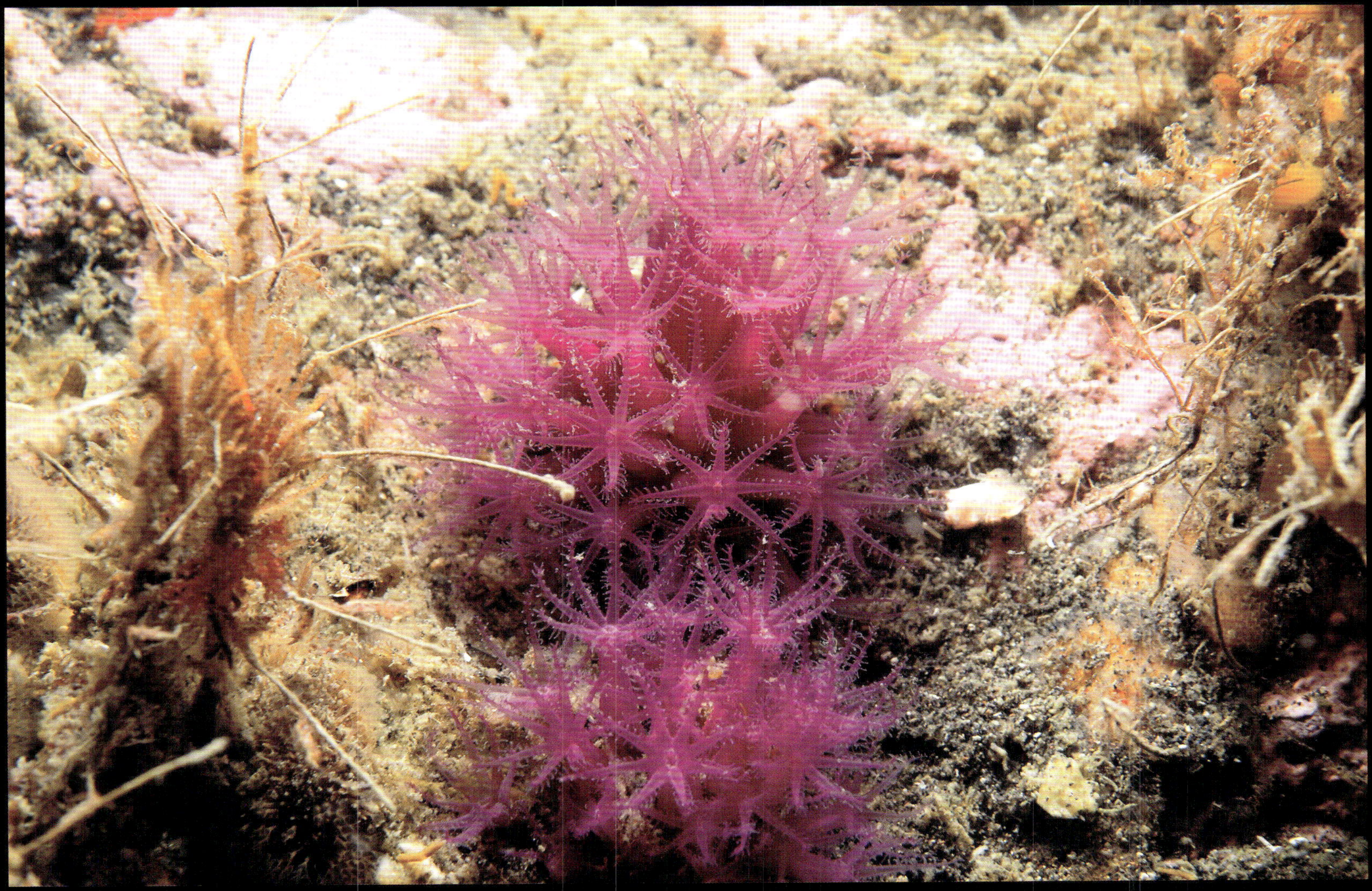

Soft corals (*Anthomastus* sp.) are a rare sight in the Aleutian Islands nearshore. When fully expanded the polyps look like hands waiting for their next meal to drift by. When the polyps retract, the colony becomes a fraction of its size as almost all of the water in the polyps is expelled. Soft corals are related to jellyfish, sea anemones, and sea pens. *Unalaska Island, S. Harper photo.*

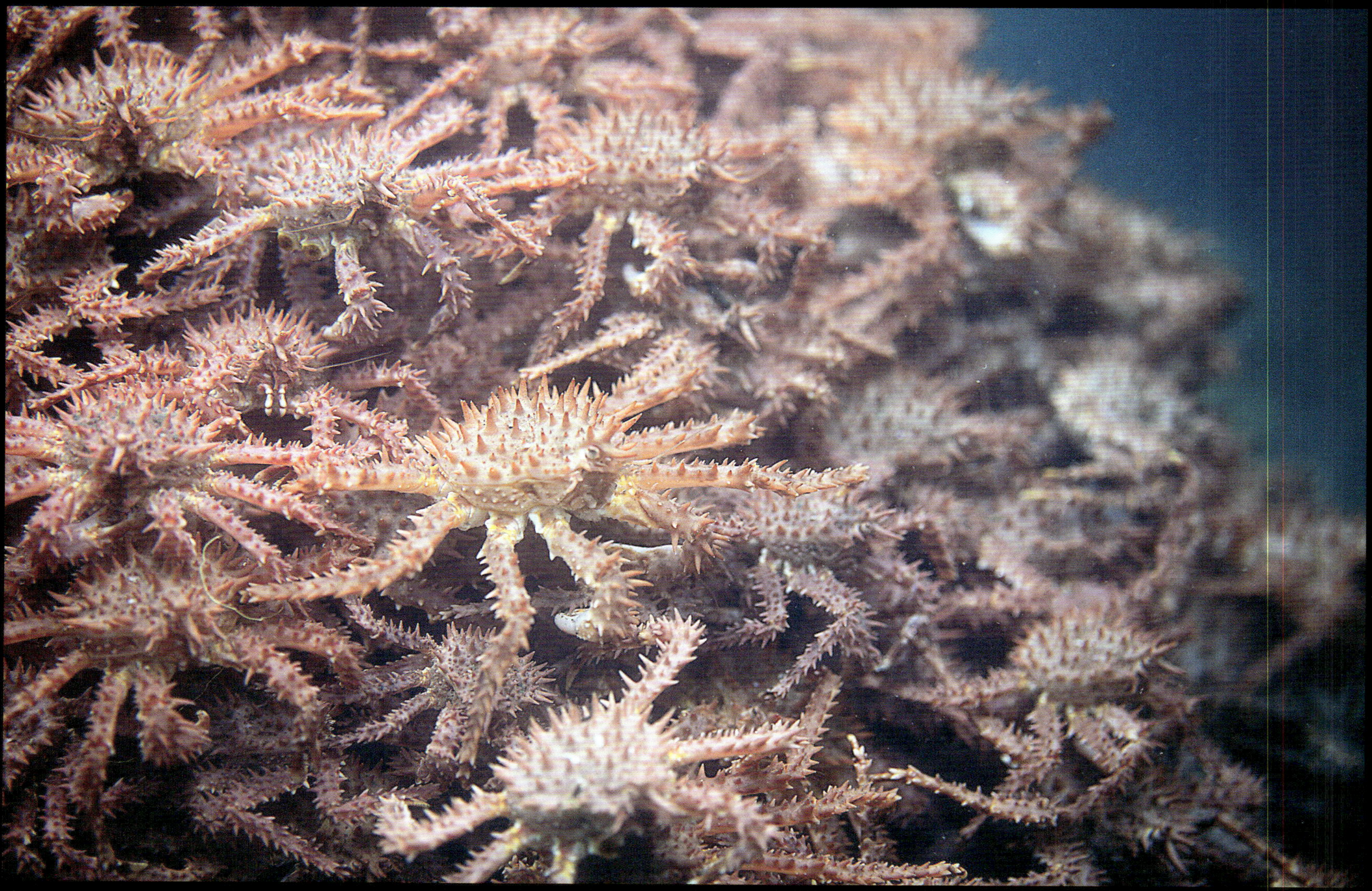

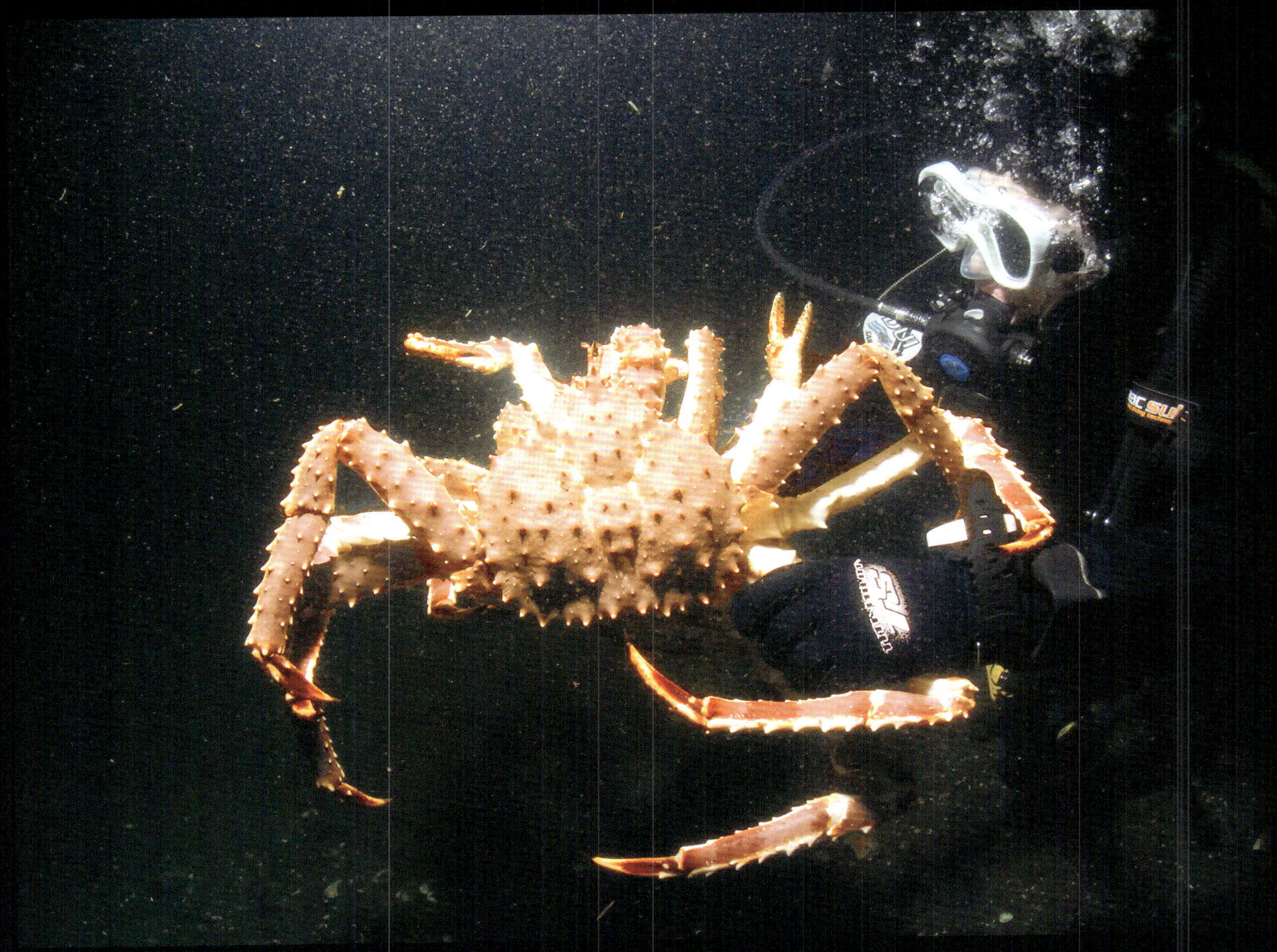

Red king crab (*Paralithodes camtschaticus*) can be difficult to manage while scuba diving, as shown here. Adults of this species can grow as large as 26 lbs (12 kg) along the Alaska coastline. This important fishery resource feeds on a variety of invertebrates including sea stars, urchins, clams, and barnacles. Juvenile red king crab are sometimes found in congregations called pods. *Left: Unalaska Bay, Unalaska Island, R. Brewer photo. Right: Nateekin Bay, Unalaska Island, S. Harper photo.*

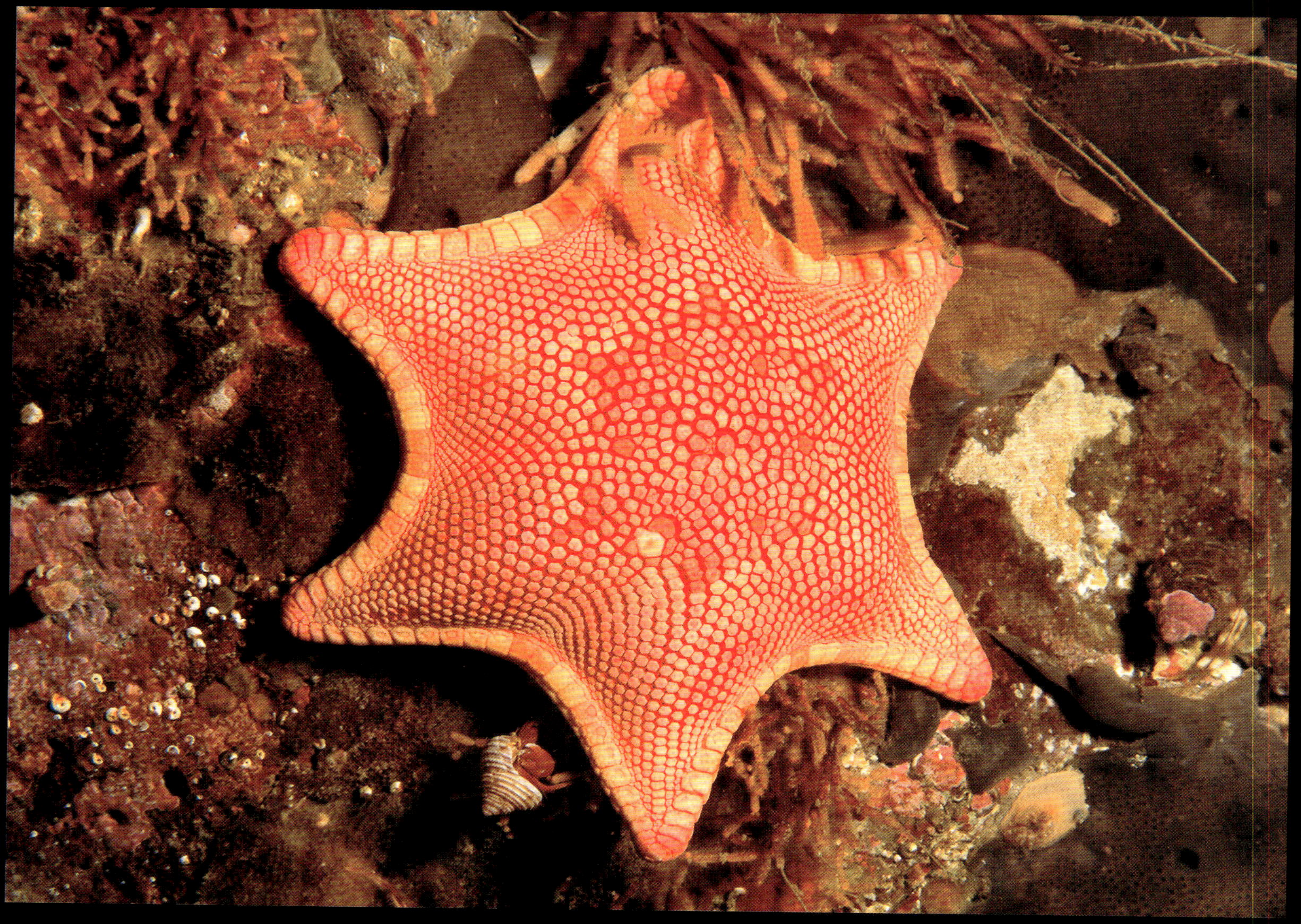

The arctic cookie star (*Ceramaster arcticus*) moves across some compound ascidians. This species has variable color morphs and is identifiable by short, triangular rays (arms) and a lack of aboral (back) spines. The arctic cookie star typically has five arms, but occasionally individuals have six arms like the one in this photo. *Ceramaster arcticus* is found from the Bering Sea to northern British Columbia, from the low tide line down to 610 ft (186 m). *Usof Bay, Unalaska Island, S. Harper photo.*

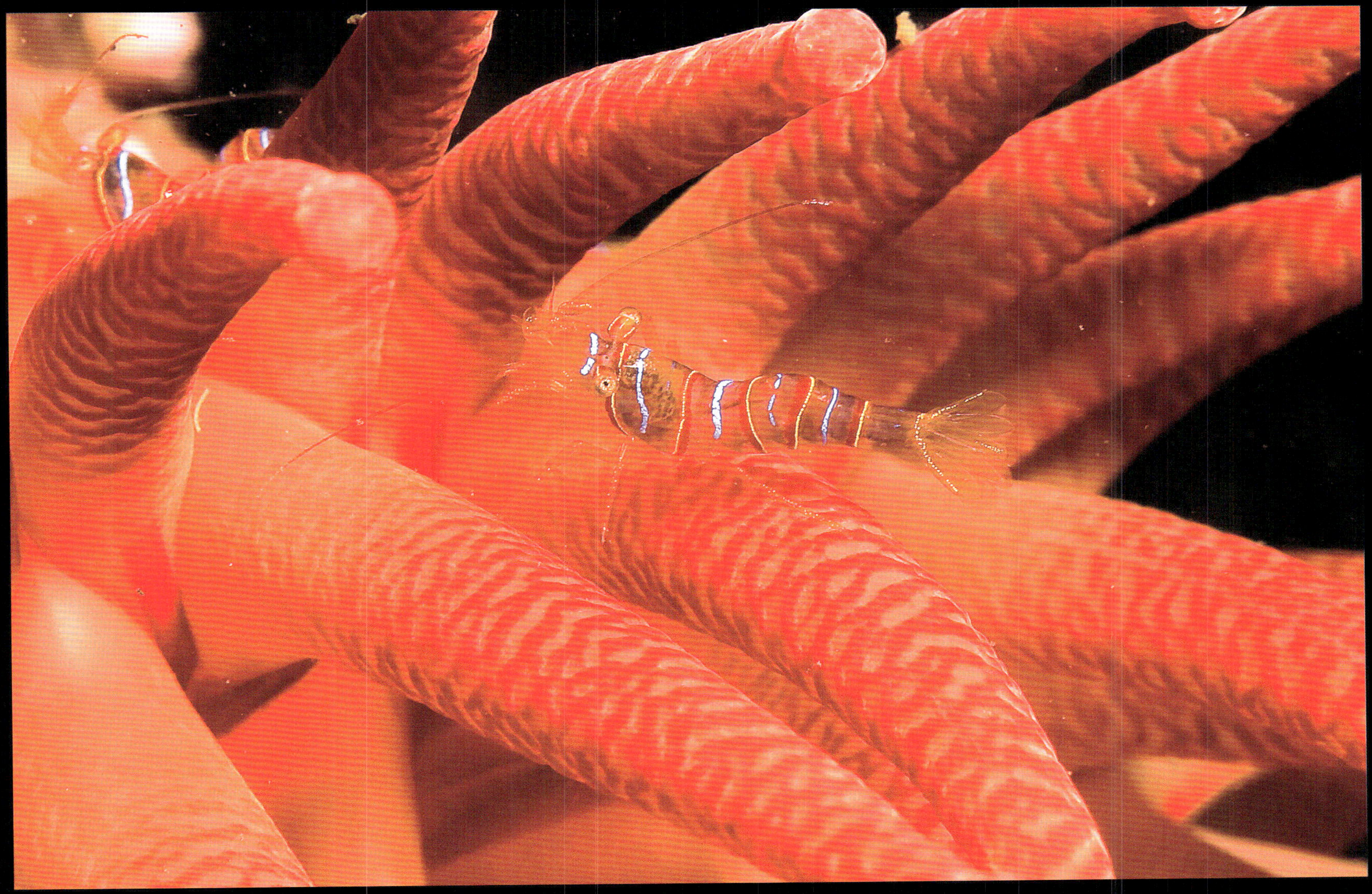

The candy stripe shrimp (*Lebbeus grandimanus*) is usually associated with the crimson anemone (*Cribrinopsis fernaldi*), shown here. This distinctive shrimp is identifiable by its brilliant bands of red, yellow, and blue and occurs from the Bering Sea to Puget Sound at depths of 20-590 ft (6-180 m). *Kismaliuk Bay, Unalaska Island, S. Harper photo.*

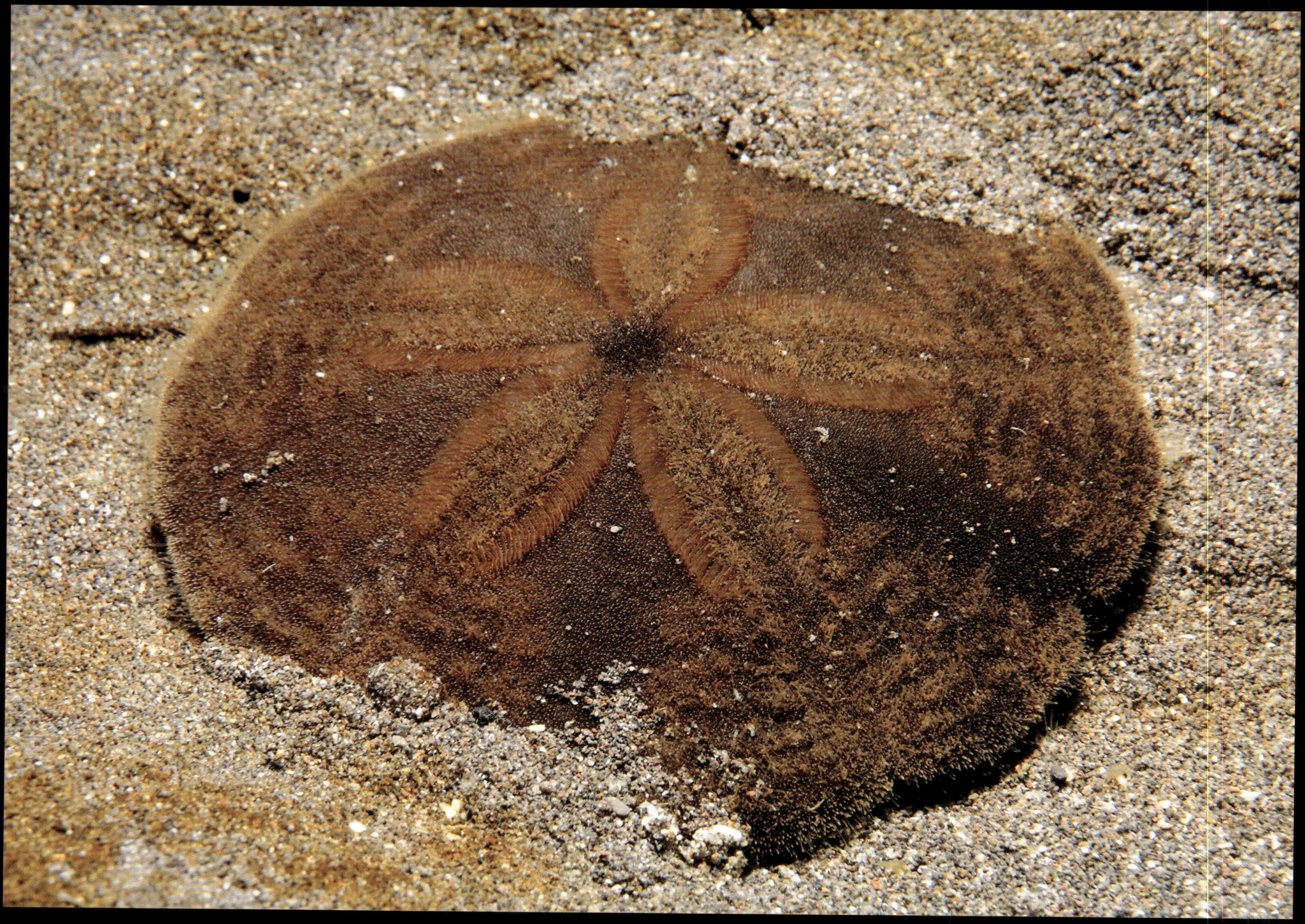

The northern sand dollar (*Echinarachnius parma*) maneuvers through sand using tiny tubefeet. Sand dollars are close cousins to sea urchins. The smooth white tests (shells) are a favorite beachcombing treasure. *Echinarachnius parma* has a broad distribution: eastern North America, northeast Asia, and western North America from Point Barrow to Prince William Sound. The depth range is from intertidal to 500 ft (150 m). *Kalekta Bay, Unalaska Island, S. Harper photo.*

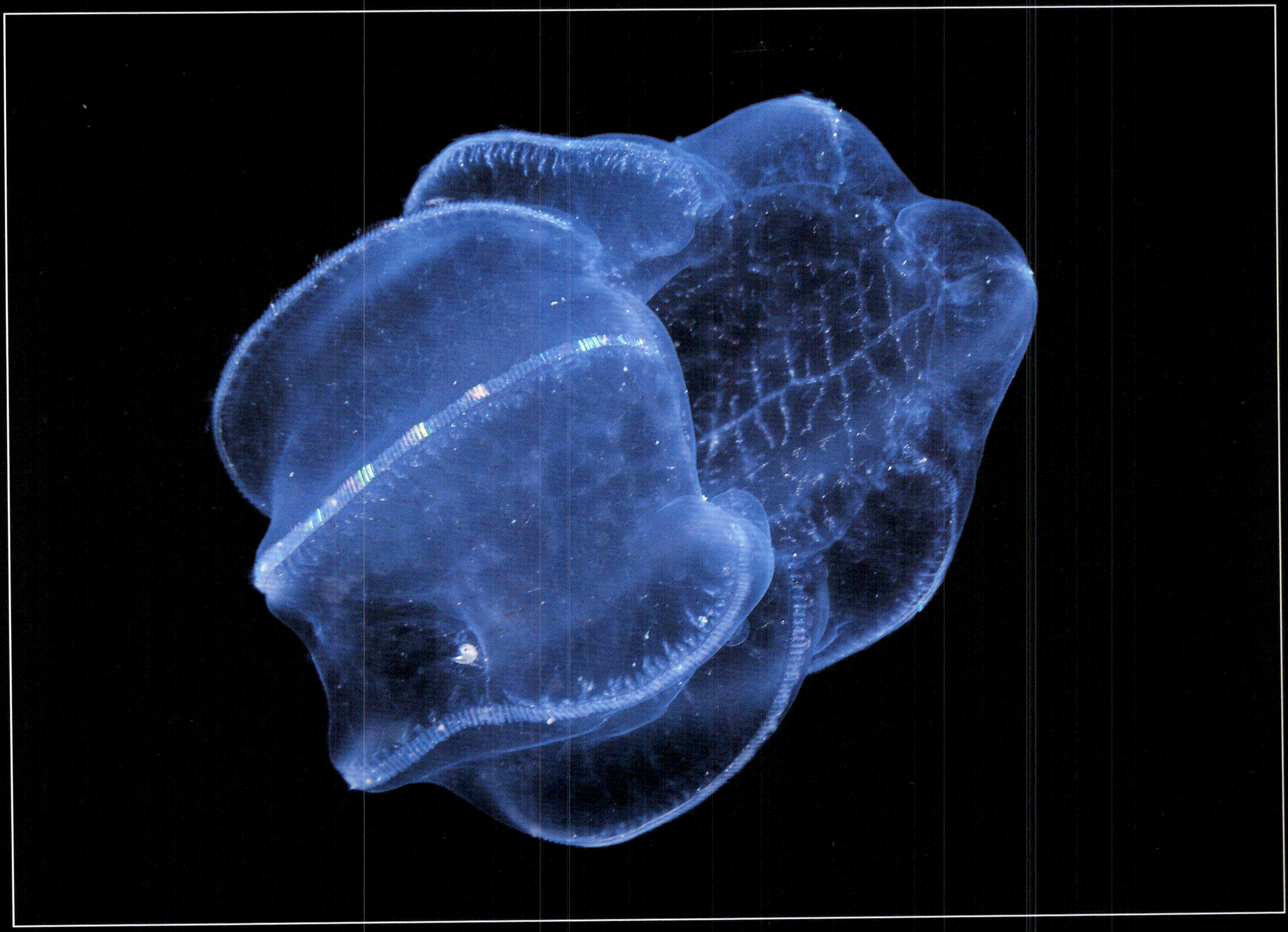

Ctenophores, also know as comb jellies, are voracious predators on planktonic (drifting) prey such as copepods and euphausiids. Ctenophores, like this 6 in (15 cm) long *Beroe* sp., are distinguished from jellyfish by their eight rows of tiny cilia (hairs) called ctenes, which are used for locomotion. A beautiful shimmering iridescence results when light strikes the beating cilia. Many ctenophores also can produce bioluminescent light. The twenty-plus species of ctenophores found from Alaska to California are eaten primarily by fishes and sea anemones. *Kalekta Bay, Unalaska Island, S. Harper photo.*

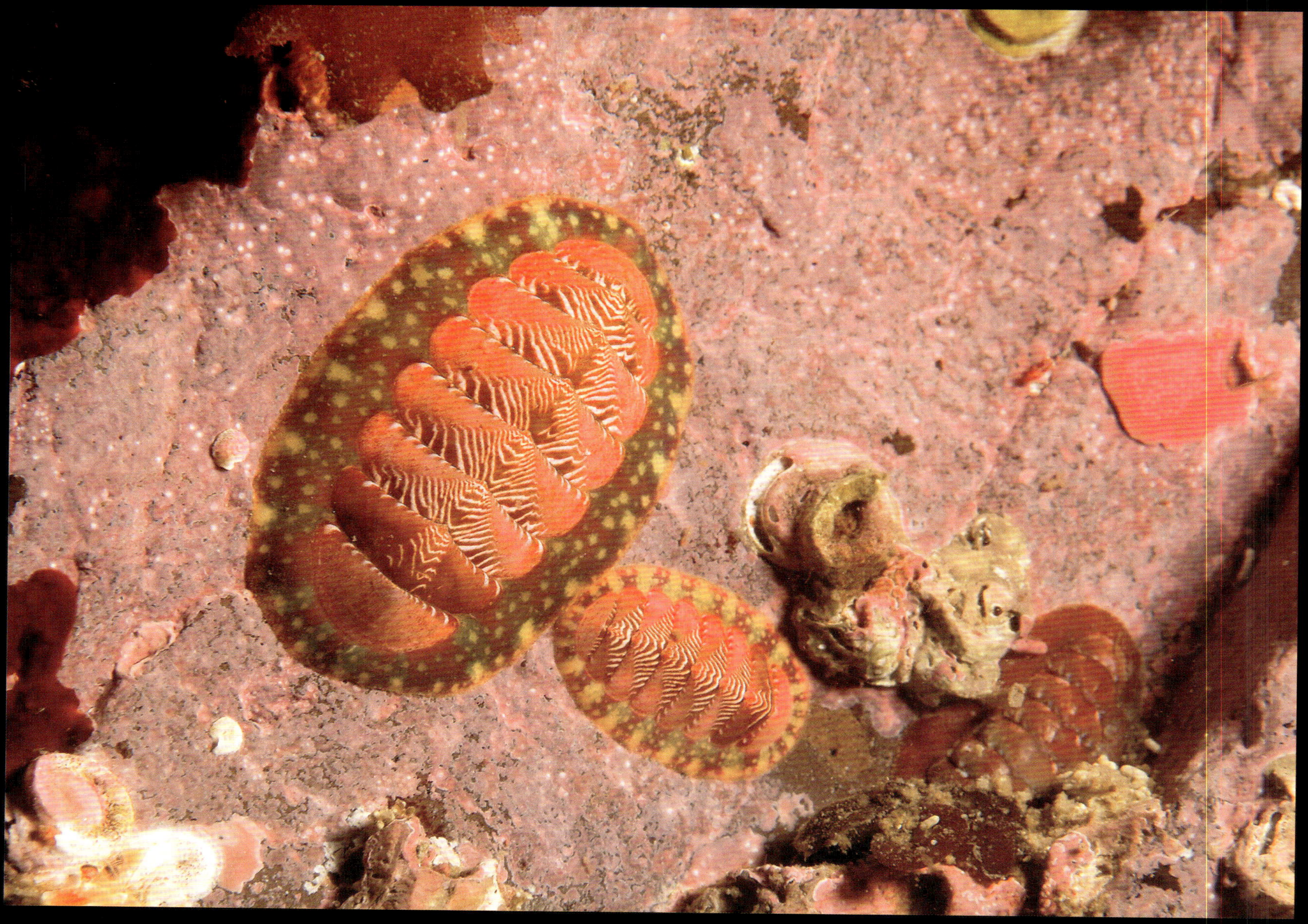

The white lined chiton (*Tonicella insignis*) has distinctive yellow spots on its mantle, and a pink to reddish body with a white triangular pattern at the apex of its valves. Chitons have eight distinctive plates or valves, which are rudimentary shells, and use a rasping tongue called a radula to scrape food from the substrate. This species is found from northern Alaska to northern Oregon. *Eagle Bay, Unalaska Island, S. Harper photo.*

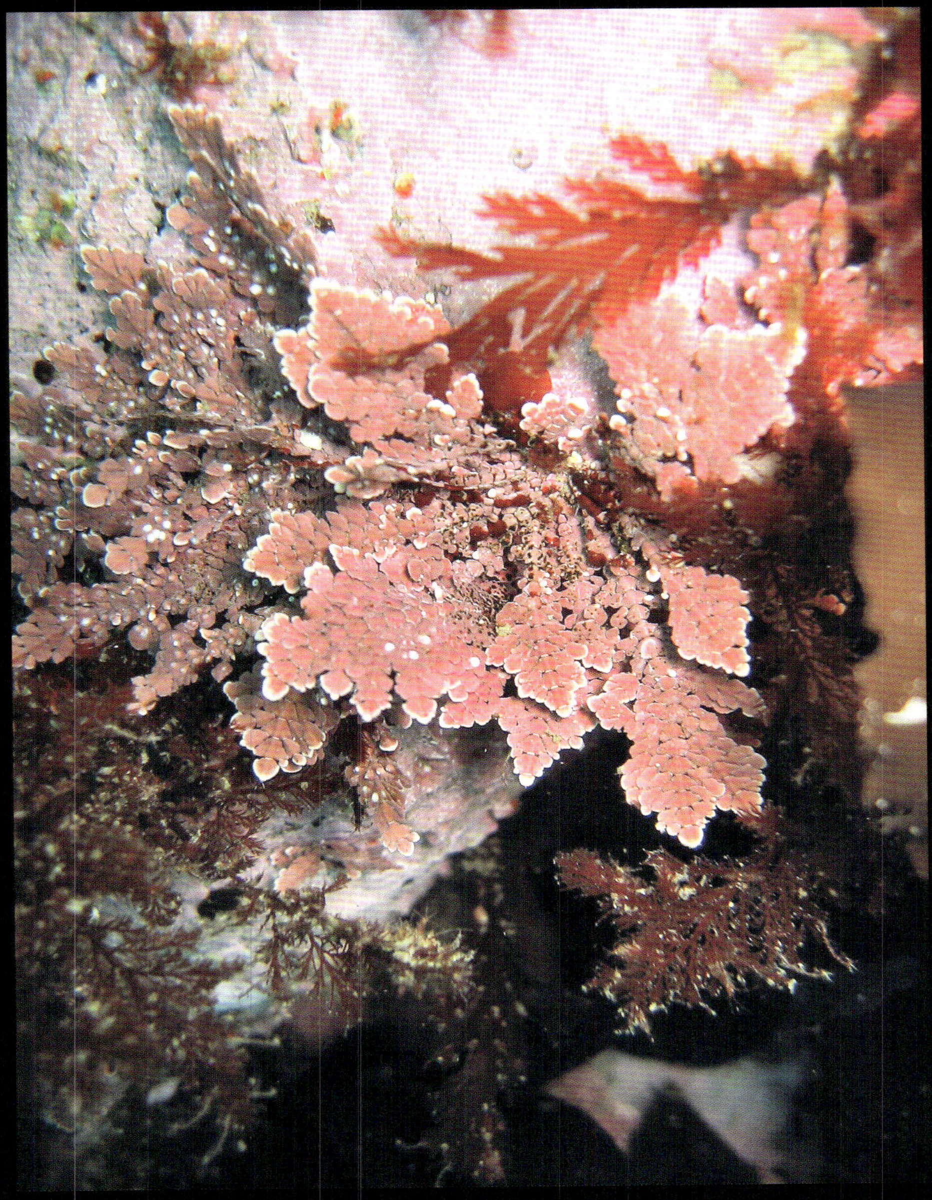

The articulated coralline red alga (*Corallina frondescens*) is prevalent from the low intertidal to subtidal zone. Like all coralline red algae, this species deposits calcium carbonate and carries out photosynthesis using chlorophyll. Unlike crustose corallines, articulated coralline algae are upright and often branching. *Tigalda Bay, Tigalda Island, S. Jewett photo.*

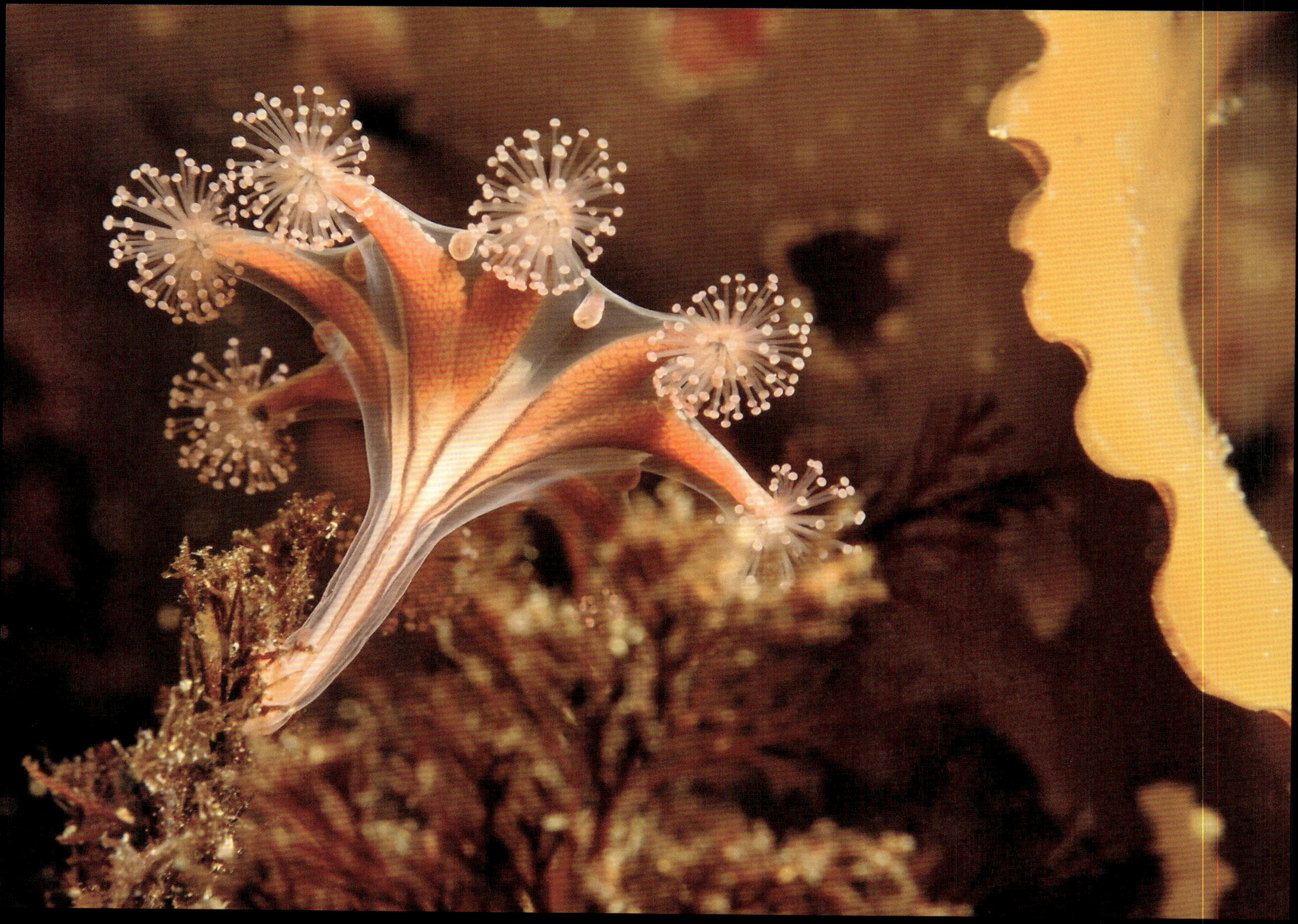

The oval-anchored stalked jelly (*Haliclystus stejnegeri*) is a jellyfish that waits for its food to pass by, rather than swim in pursuit and drift aimlessly. This species inhabits surge-swept areas and is usually found attached to algae or rocks. It captures food (mostly tiny crustaceans) using its eight multi-tentacled arms. *Cape Prominence, Unalaska Island, S. Harper photo.*

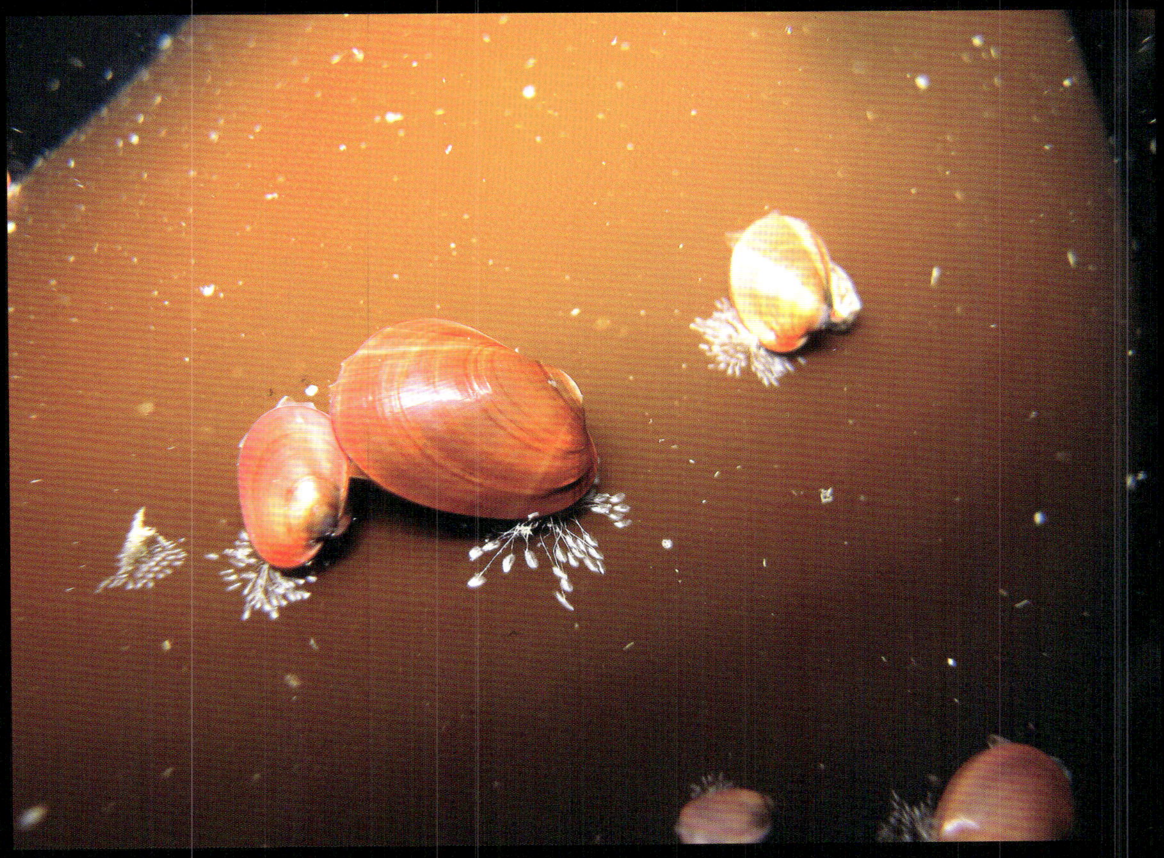

The varnished crenella mussel (*Vilasina vernicosa*), like the blue mussel (*Mytilus trossulus*), secretes strong, silken byssal threads to attach to the substrate. These mussels are attached to a brown alga (*Saccharina* sp.). *Peacock Point, Unalaska Island, S. Jewett photo.*

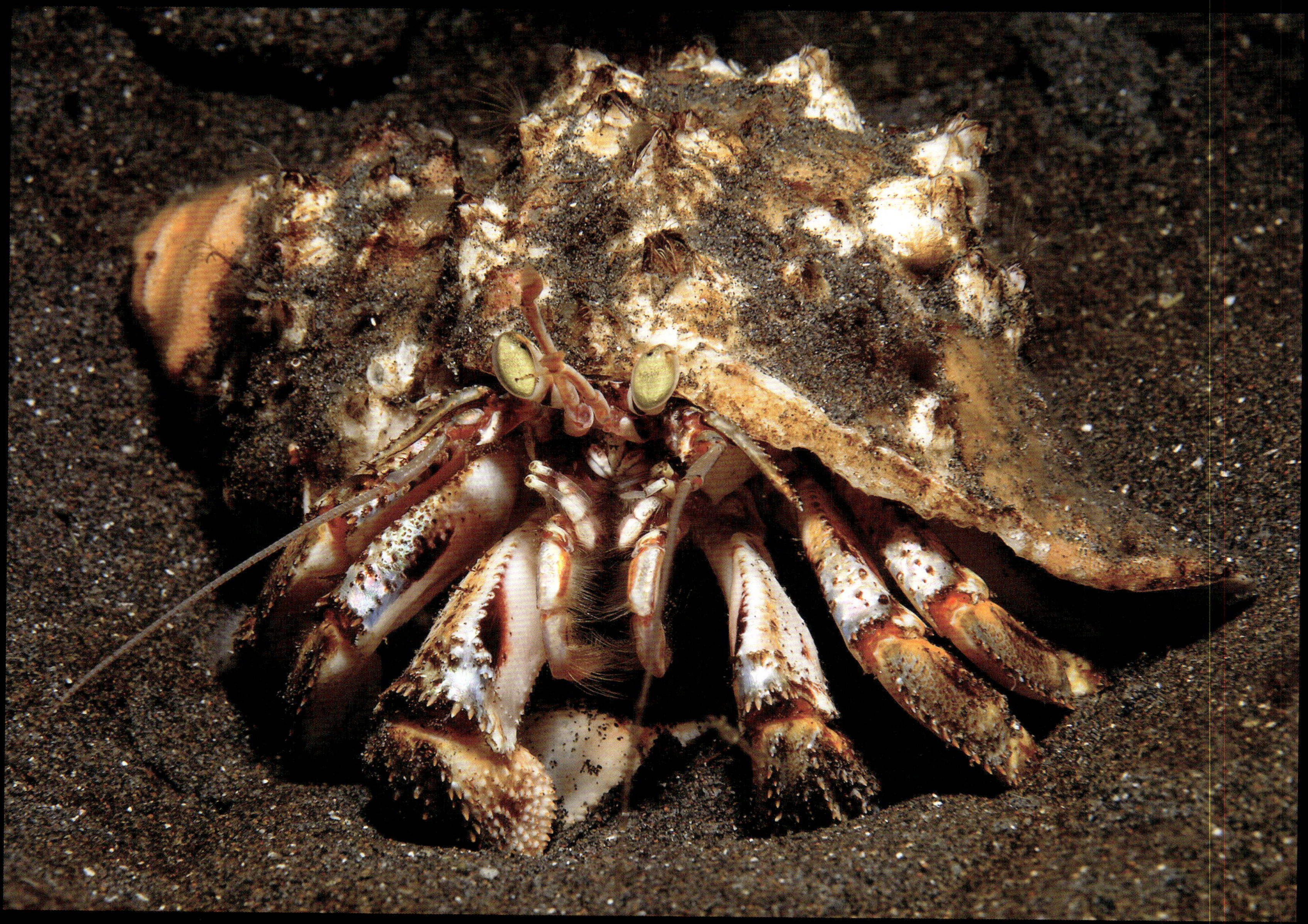

The Alaskan hermit crab (*Pagurus ochotensis*) has distinctive yellow eyes and red, white, and blue markings on its legs. When approached, this shy species will often run for several meters before collapsing into its shell. *Kisselen Bay, Unalaska Island, S. Harper photo.*

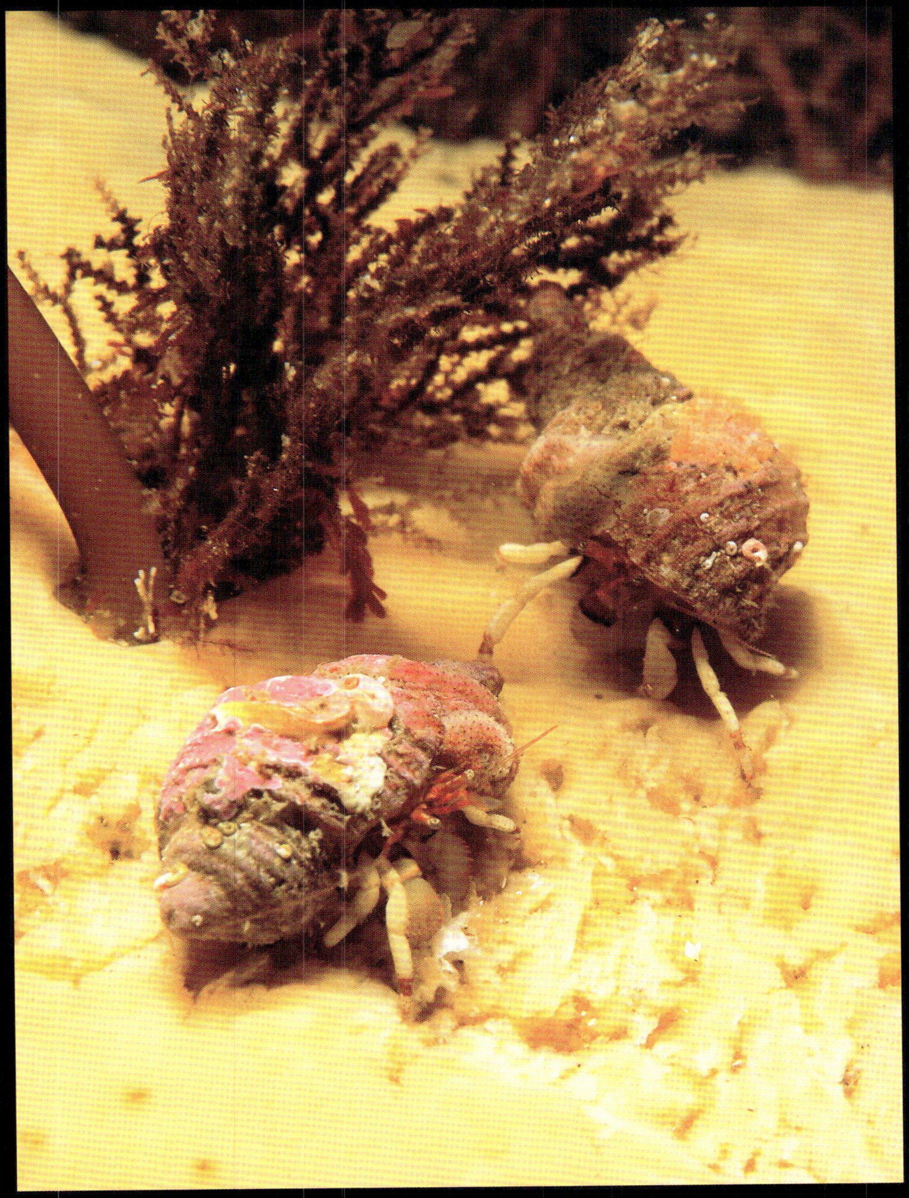

Two hermit crabs (Paguridae) square off on a breadcrumb sponge. Most species of hermit crabs are identified by banding or colors on their legs. Like all crustaceans, hermit crabs molt (shed their exoskeleton) as they grow. When a hermit crab outgrows its shell-home it must find another. *Lookout Point, Umnak Island, S. Harper photo.*

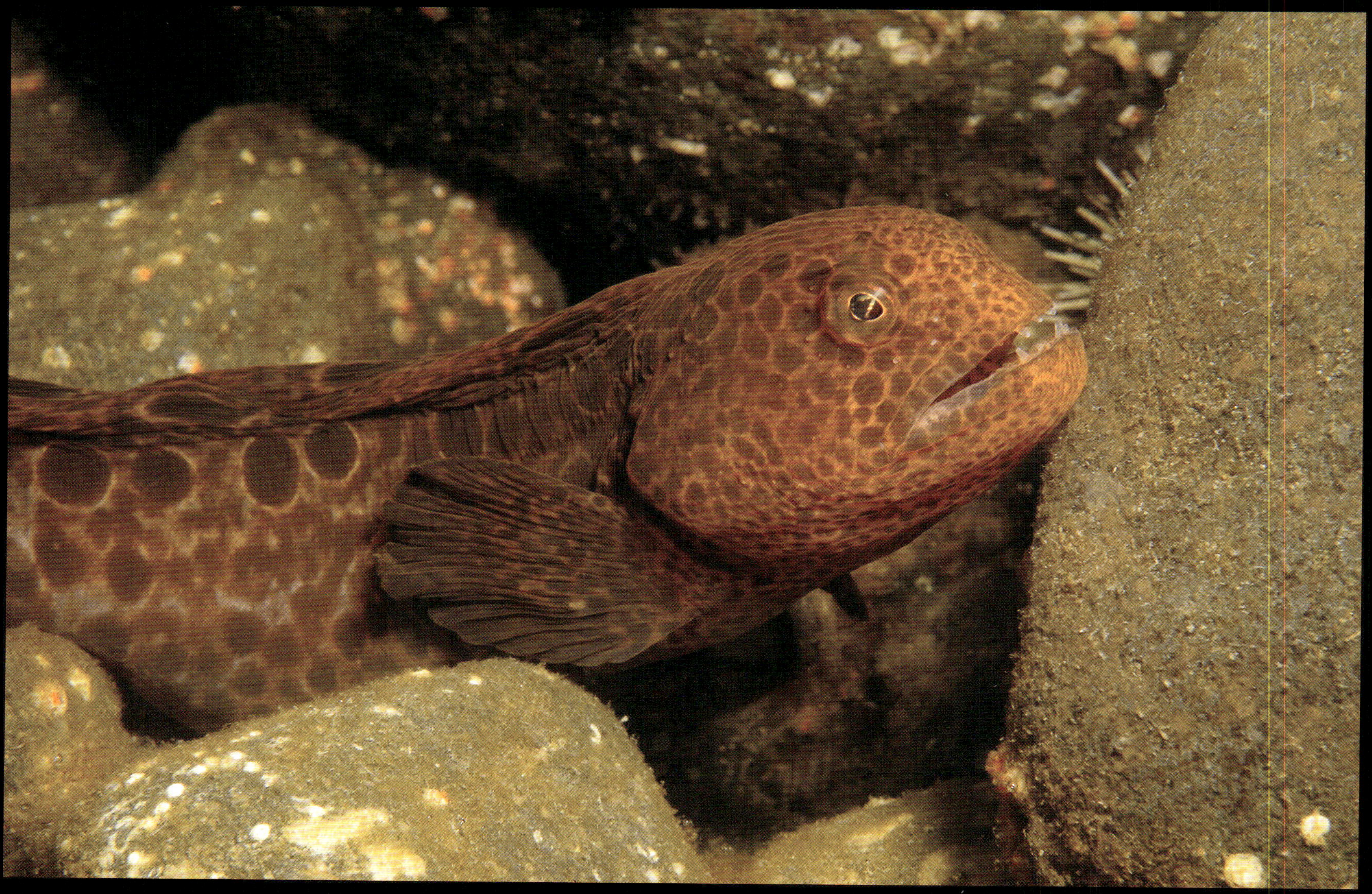

The wolf-eel (*Anarrhichthys ocellatus*) can reach 8 ft (2.5 m) in length and is identified by large puffy lips, canine-like teeth, and prominent black spots on most of the gray body. This fish has flattened teeth in the back of its mouth that allow it to eat hard-shelled prey like sea urchins, crabs, clams, mussels, scallops, and snails. Juveniles, like the one shown here, have maroon coloration. *Kalekta Bay, Unalaska Island, S. Harper photo.*

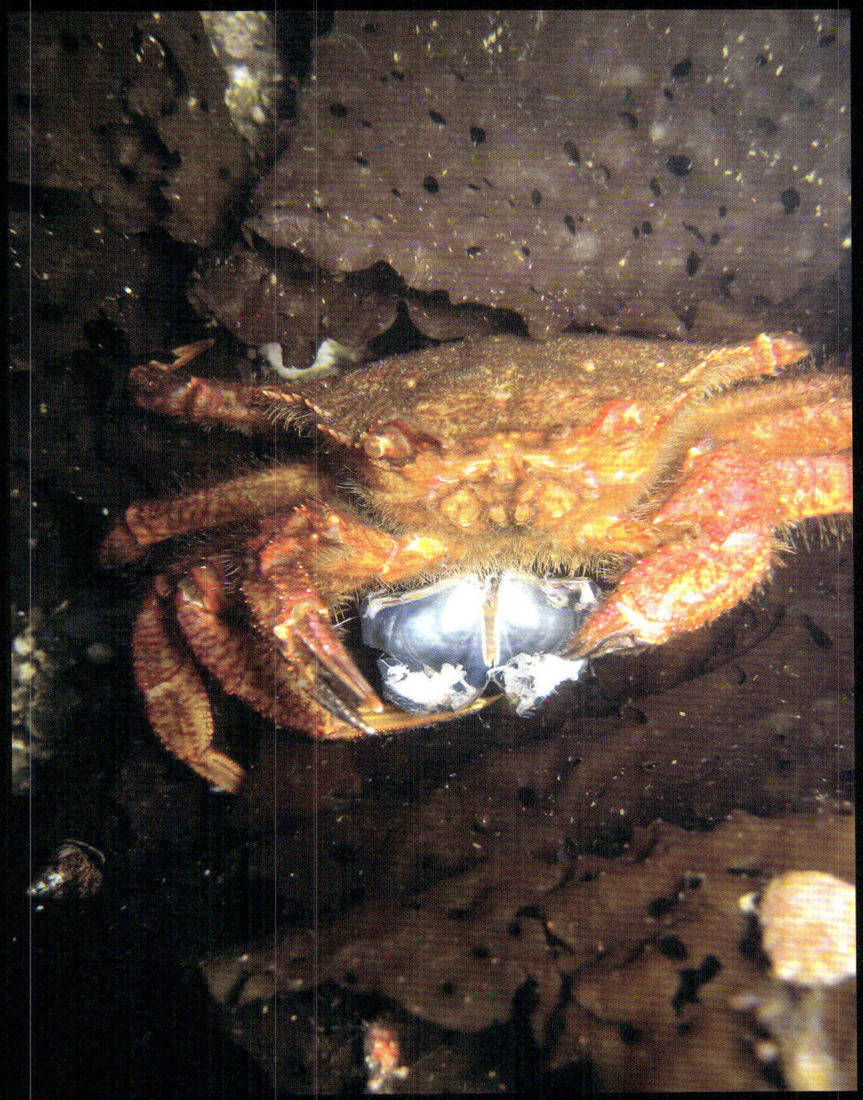

The helmet crab (*Telmessus cheiragonus*) inhabits subtidal eelgrass beds and understory kelps. This fast-moving crab will often sprint away when a diver approaches. Population estimates and life-history information for helmet crabs are difficult to get because they can easily bury in the sand, particularly females. This ubiquitous crab occurs on varied substrates mainly in nearshore waters from the northwest Chukchi Sea to northern California. It is consumed by several predators including sea otters, bearded seals, and fishes. The crab in this photo is eating a blue mussel. *Kismaliuk Bay, Unalaska Island, R. Brewer photo.*

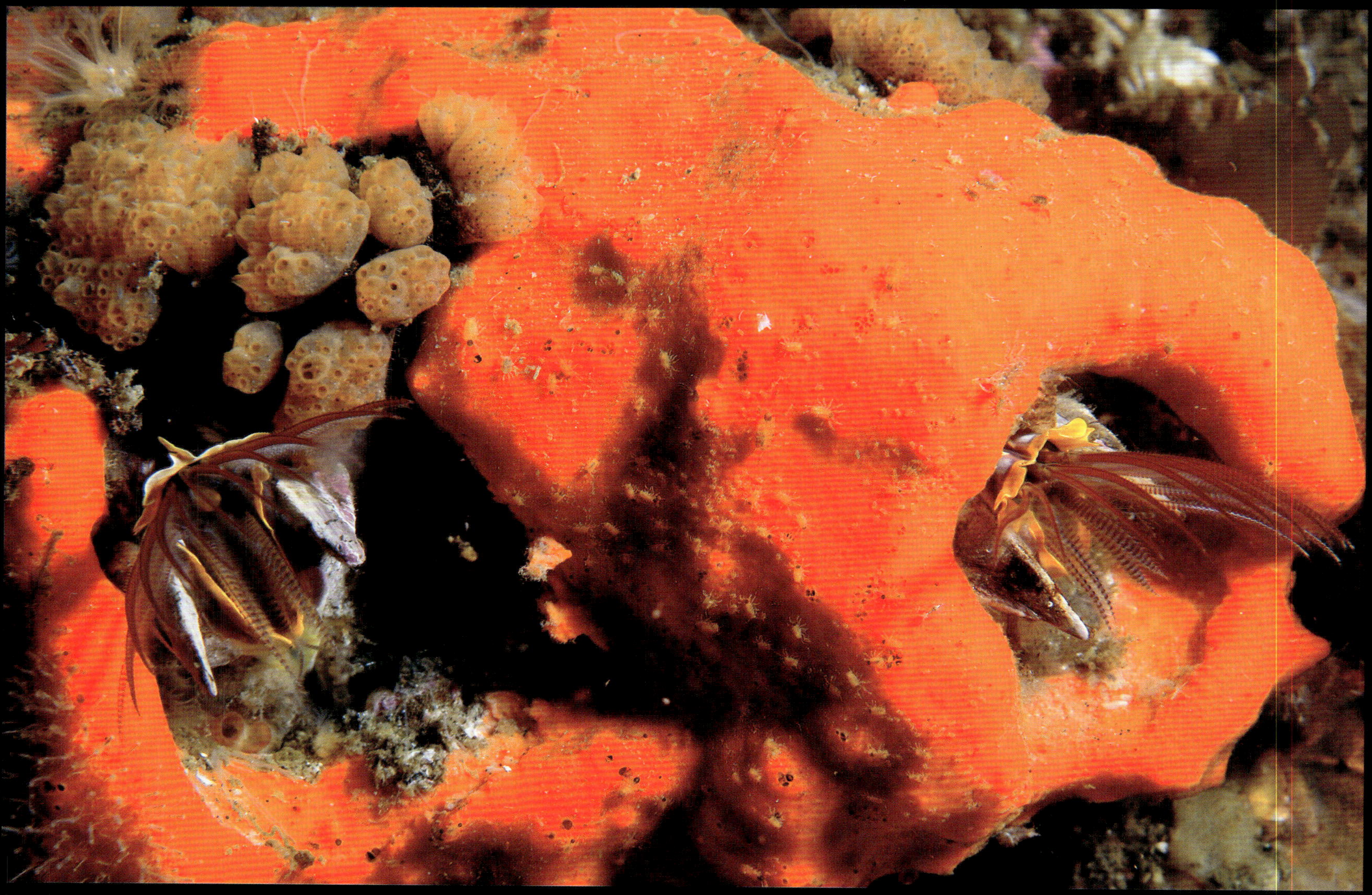

The giant acorn barnacle (*Balanus nubilus*) is identified by the gold markings around the cirri (feeding appendages). The barnacle may be solitary or in aggregate, and the empty casings found after the barnacles have perished often provide shelter for other organisms. *Driftwood Bay, Unalaska Island, S. Harper photo.*

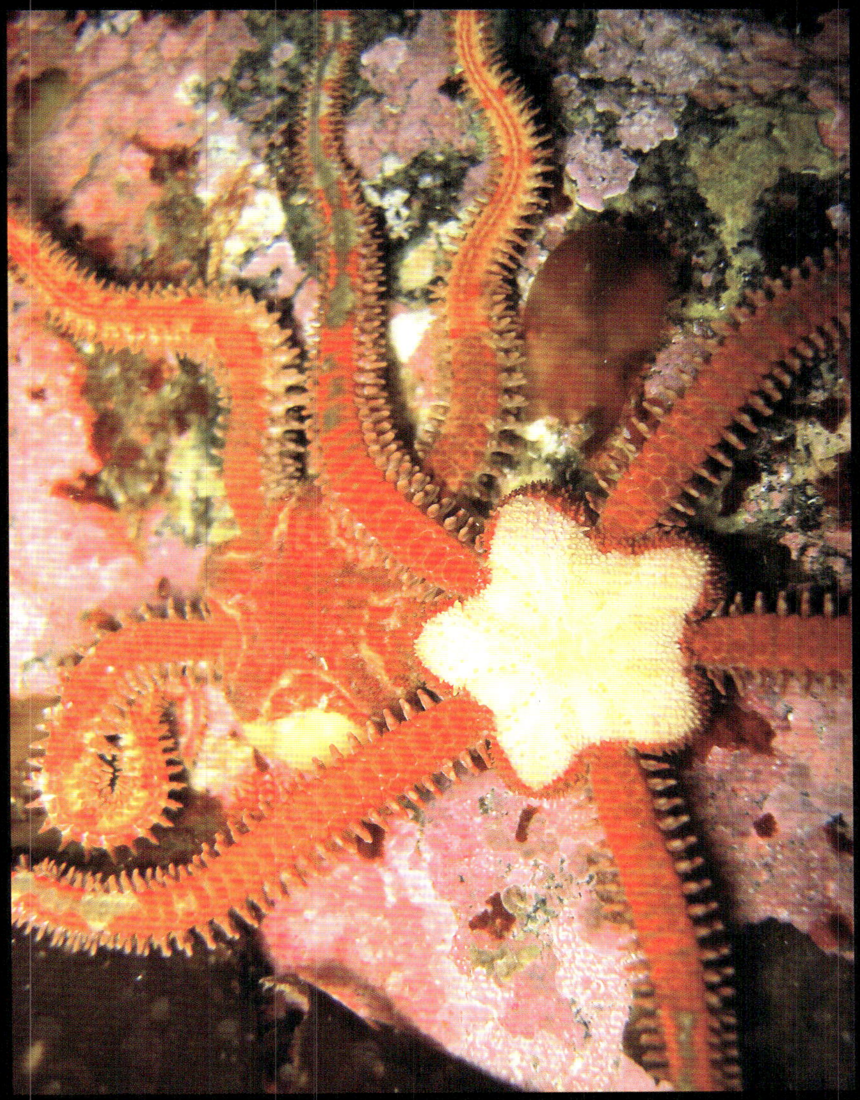

The daisy brittle star (*Ophiopholis aculeata*) is aptly named for the brittleness of its arms, which can break off when it is disturbed. Most brittle stars can survive with only one arm remaining and can grow new arms at a rate of 0.1 in (2.3 mm) per month. Usually only the tips of their arms are visible, reaching from under rocky outcroppings to catch food. Daisy brittle stars have extremely variable color patterns and are found from the low intertidal zone to a depth of 6,600 ft (2,000 m). *Peacock Point, Unalaska Island, R. Brewer photo.*

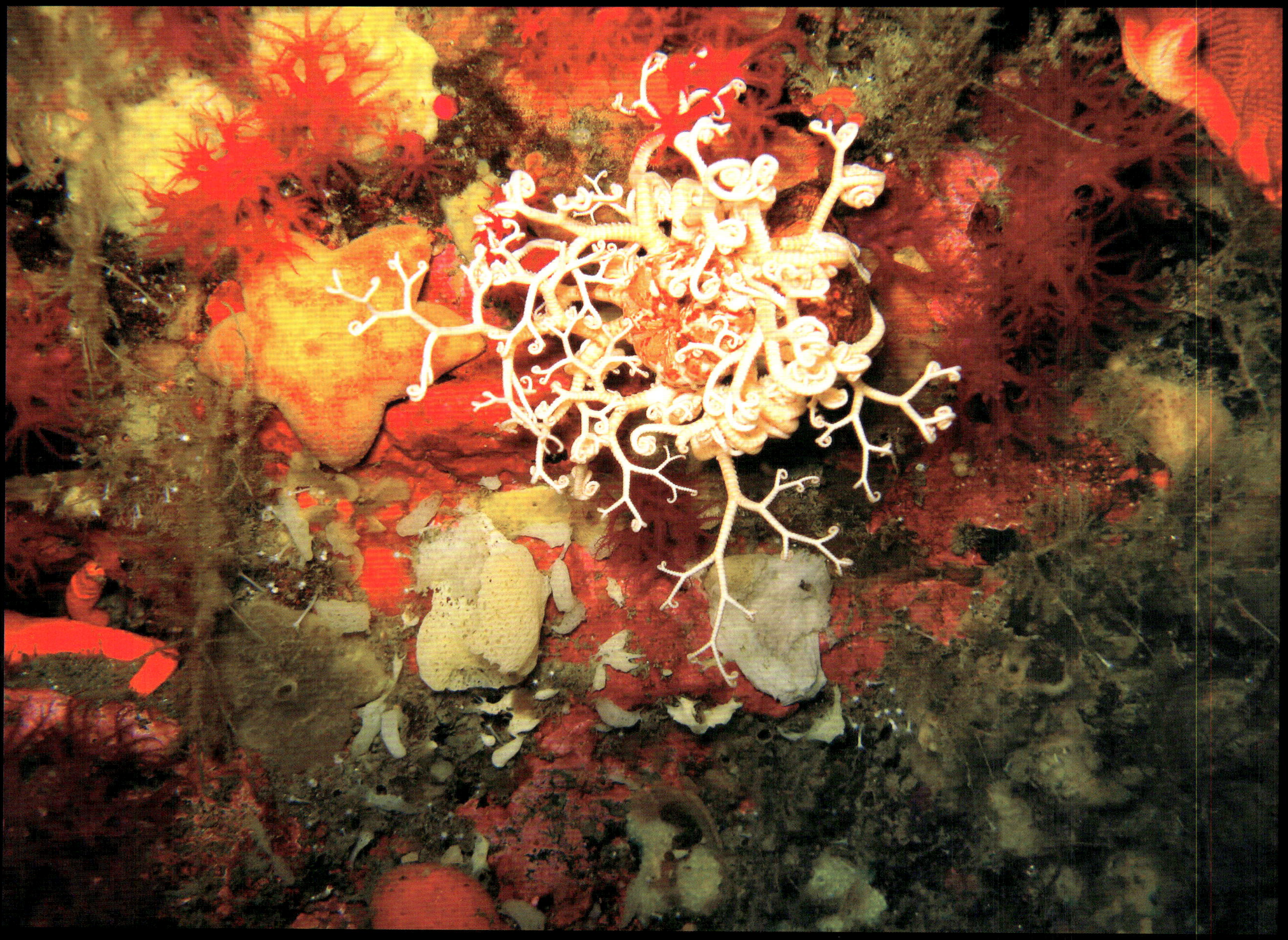

The basket star (*Gorgonocephalus eucnemis*) has five arms that branch repeatedly. During feeding the branches form a basket-like cage to catch planktonic prey. The basket star occurs throughout the Aleutians, and is found at 30–6,000 ft (6-1,850 m), but is more common below 100 ft. *Avatanak Island, R.N. Clark photo.*

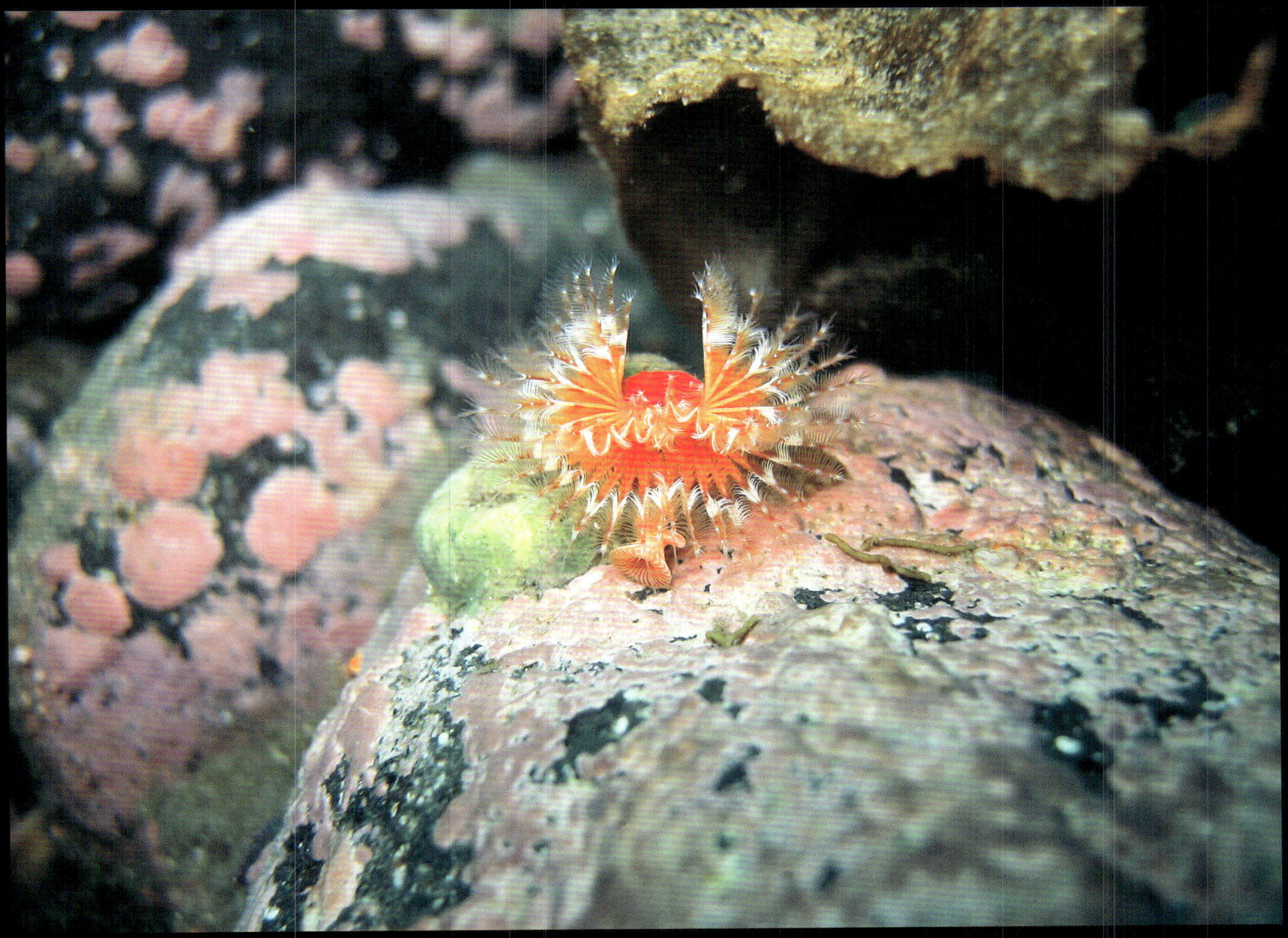

The yoke-bearer calcareous tubeworm (*Crucigera zygophora*) has a distinctive red and white corona (crown) and a characteristic operculum (trap door). Like other calcareous tubeworms, *C. zygophora* prefers hard rocky substrates and forms a calcareous tube with calcium carbonate and other secreted materials. *Eider Point, Unalaska Island, S. Jewett photo.*

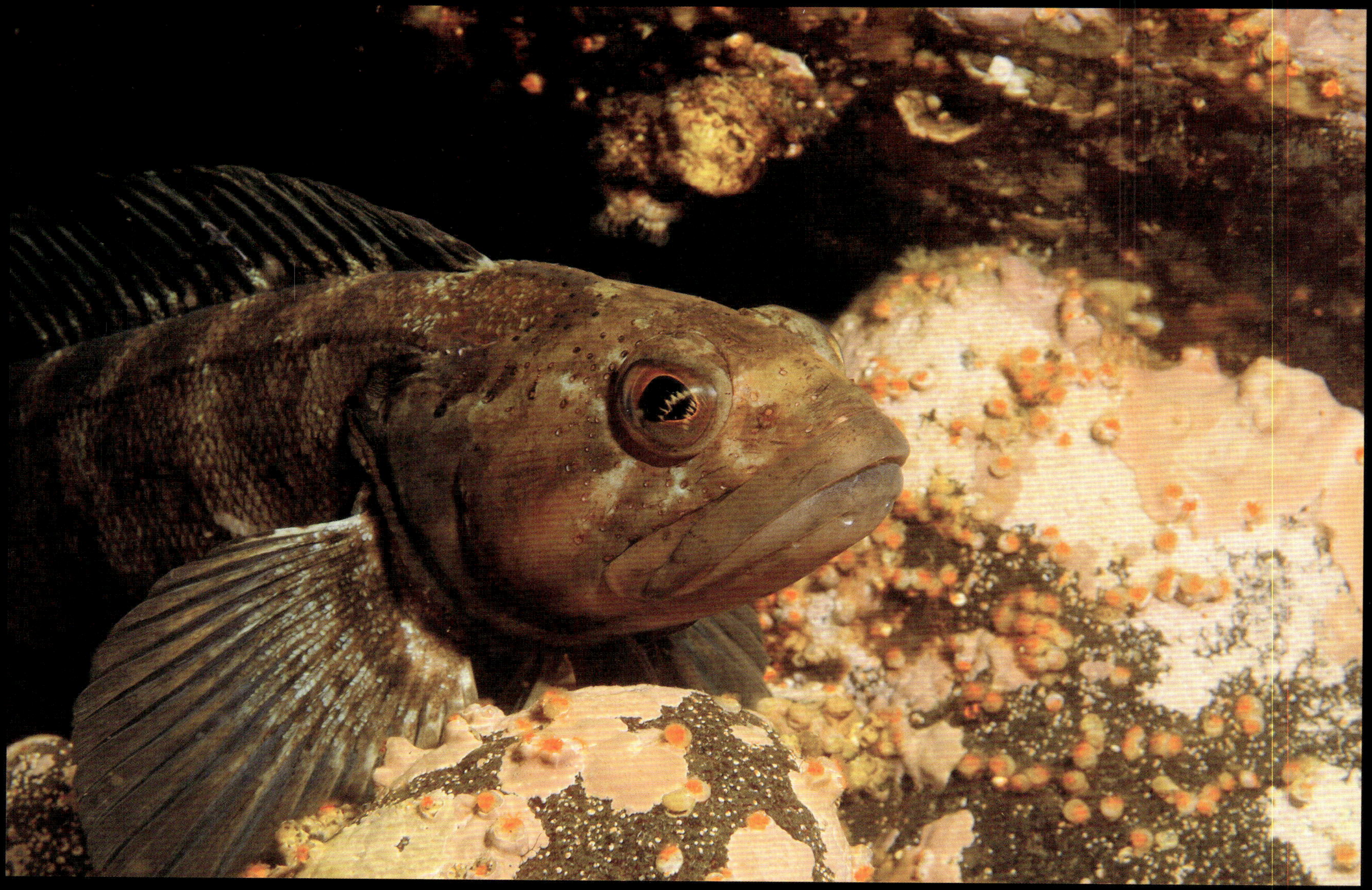

The Alaskan ronquil (*Bathymaster caeruleofasciatus*) never strays too far from its rocky cave. This curious fish is elongate with very long dorsal and anal fins composed almost entirely of soft rays. The nine species in this family occur only in the North Pacific as bottom dwellers. *Kalekta Bay, Unalaska Island, S. Harper photo.*

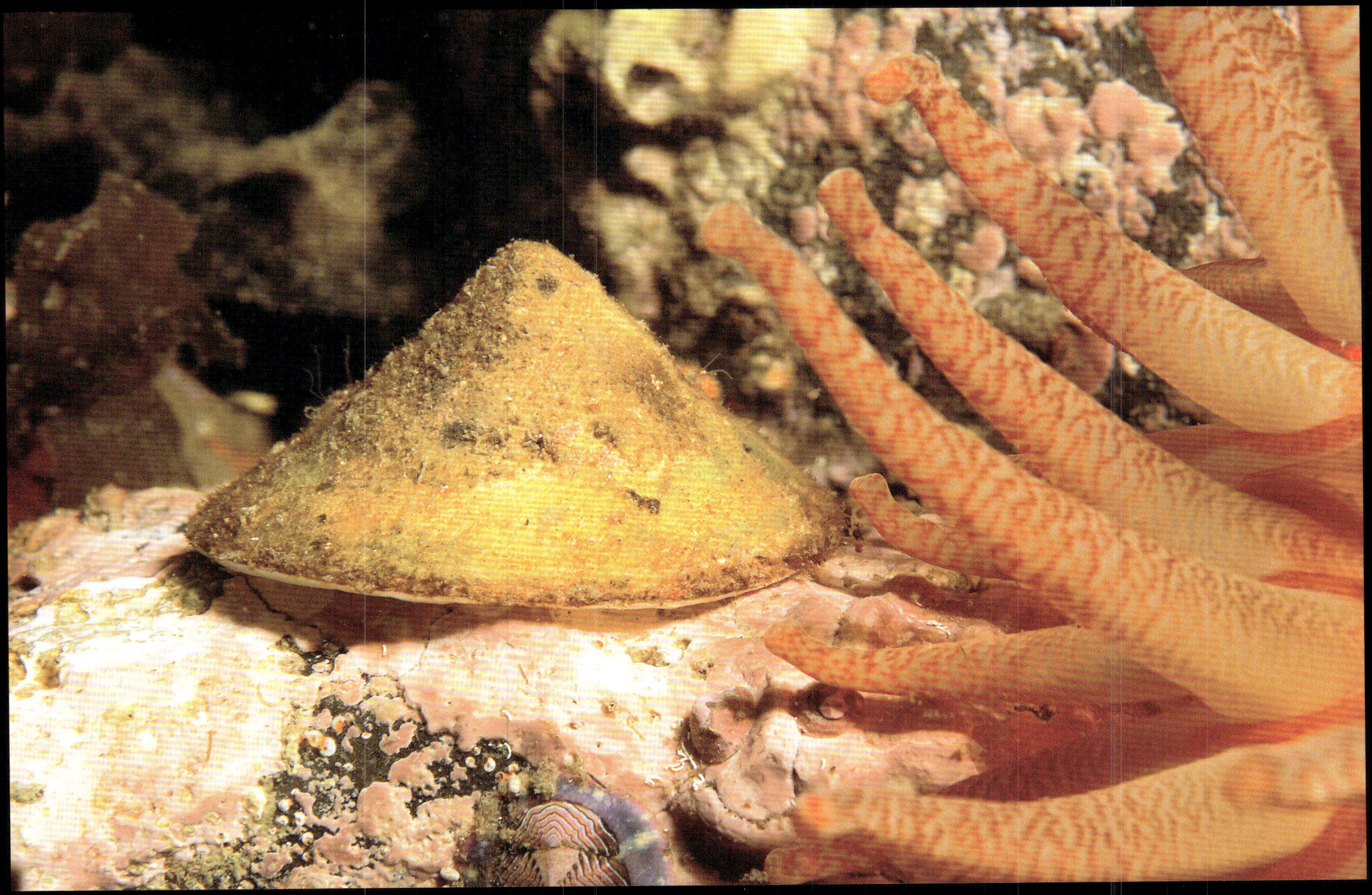

The whitecap limpet (*Acmaea mitra*) is covered with red crustose coralline algae, the very species that it feeds on. The whitecap limpet grows to 1.5 in (3.75 cm) and is found on rocks from the low tide line to 100 ft (30 m) depth from northern Alaska to northern Mexico. *Usof Bay, Unalaska Island, S. Harper photo.*

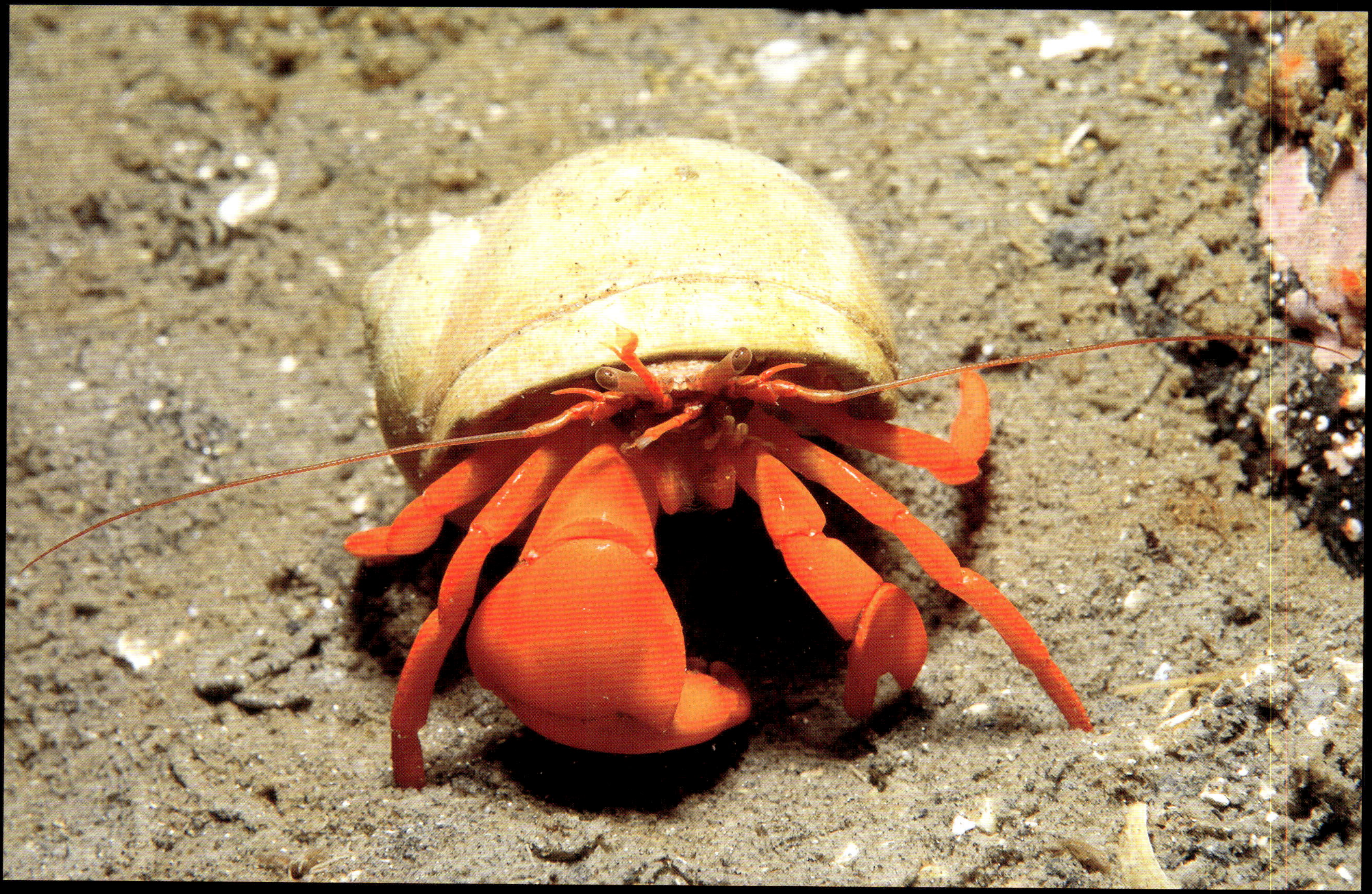

The Pacific red hermit crab (*Elassochirus gilli*) is easy to identify with its bright orange color and lack of hairs. Most hermit crabs use their large right claw to seal off the door to their home once they withdraw inside. *Elassochirus gilli* is common to depths of 650 ft (200 m) in the Bering Sea and throughout the Aleutian Islands south to northern Washington. *Makushin Bay, Unalaska Island, S. Harper photo.*

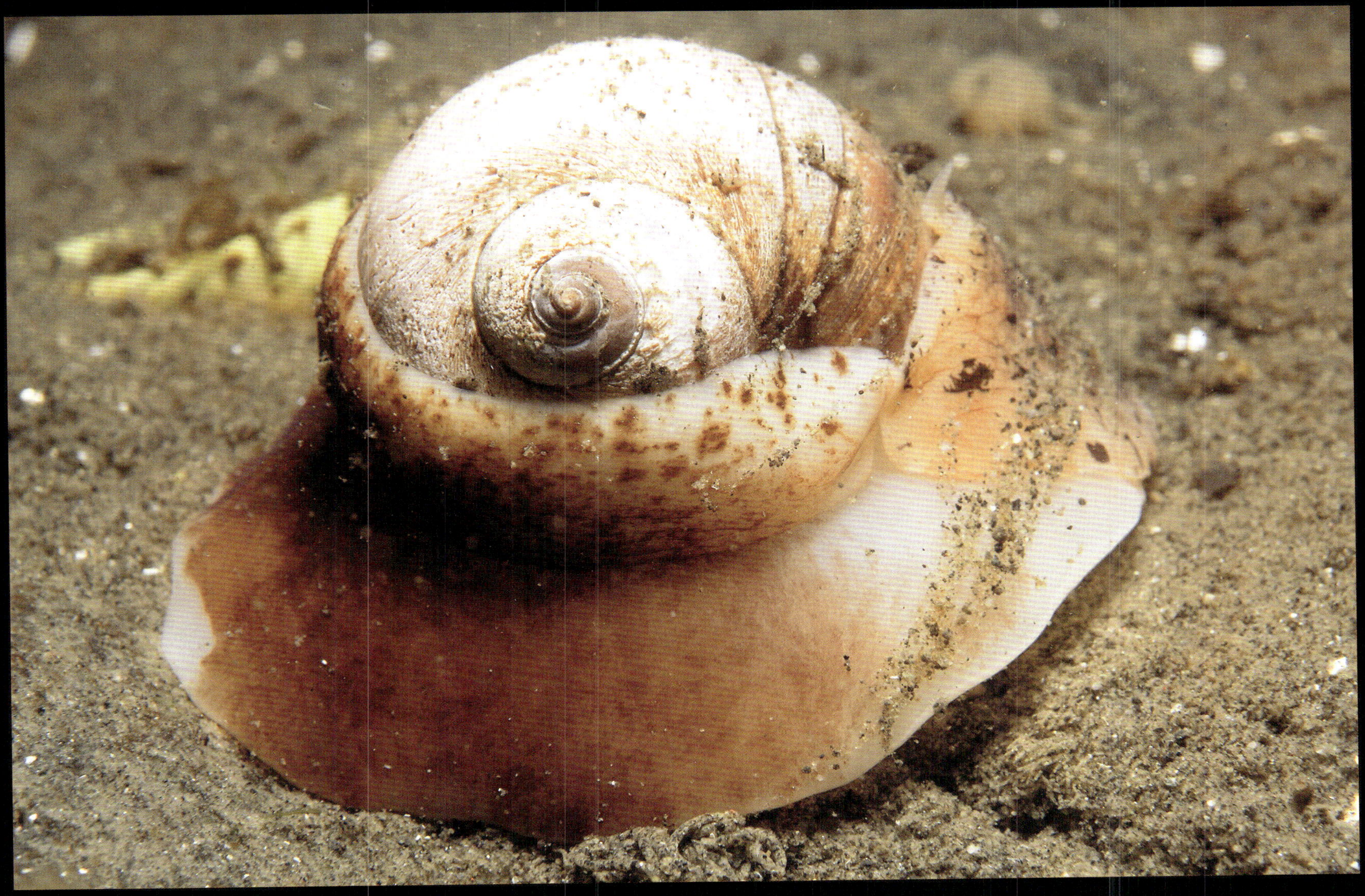

The Aleutian moon snail (*Cryptonatica aleutica*) has a distinctive moon-shaped shell and large mantle. Moon snails prey on bivalves and other snails by drilling a hole in their shells with a chitinous tongue (radula). Like other moon snails, this species burrows into soft substrates in search of food. *Makushin Bay, Unalaska Island, S. Harper photo.*

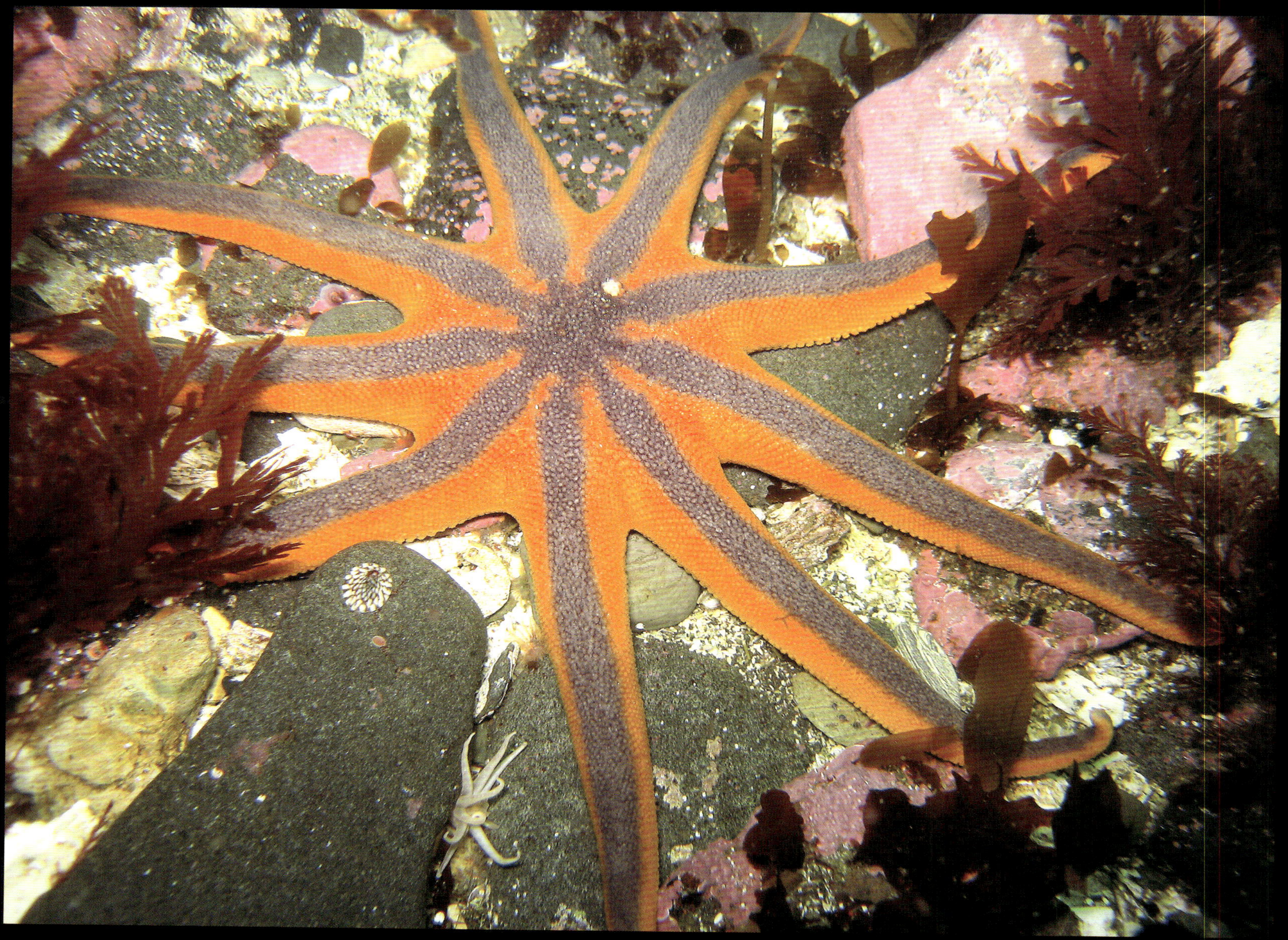

The striped sun star (*Solaster stimpsoni*) has very distinctive orange to red color with radiating gray bands. This large 9- to 10-armed star can attain diameters up to 18 in (46 cm). This species preys upon sea cucumbers, lamp shells, sea squirts, and sea pens. It ranges from Oregon to the Bering Sea to Japan and is usually associated with rocky substrates. *Nikolski Bay, Umnak Island, R. Brewer photo.*

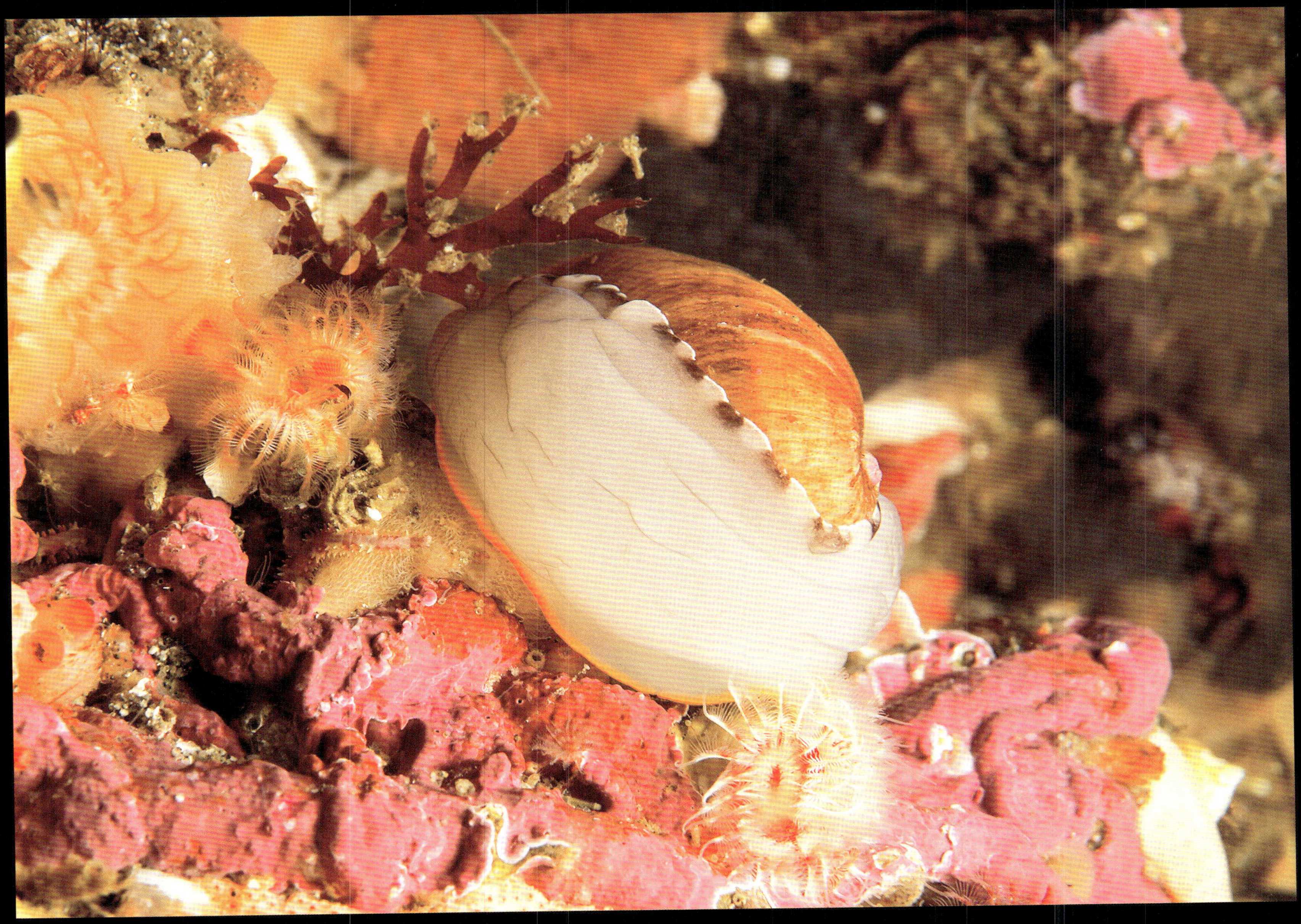

The smooth velvet snail (*Velutina prolongata*) is identified by the bright orange-red rim and banded mantle. Its shell can be nearly covered by a thin layer of tissue. *Traders Cove, Umnak Island, S. Harper photo.*

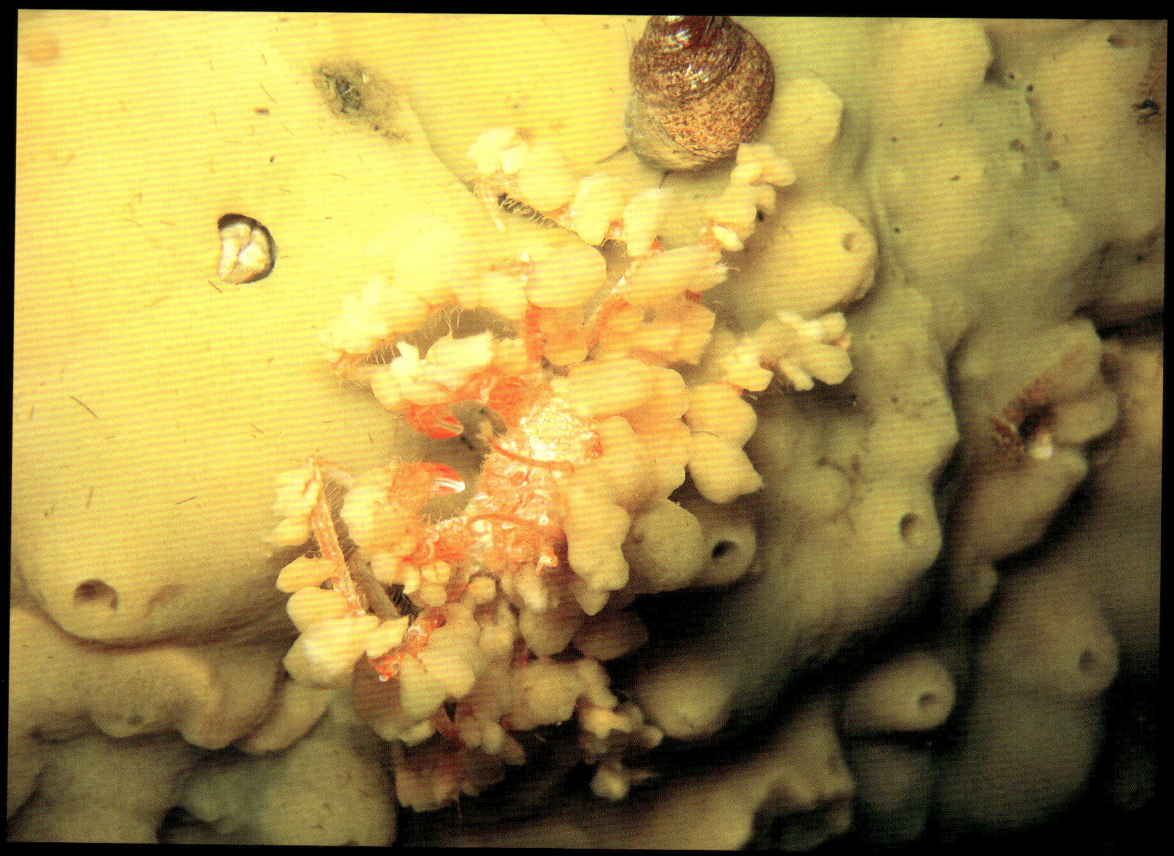

The graceful decorator crab (*Oregonia gracilis*) decorates its body with any kind of material, including marine debris. This one has an affinity for the breadcrumb sponge (*Halichondria panicea*). This species is found throughout the Aleutians. *Chernofski Bay, Unalaska Island, H. Chenelot photo.*

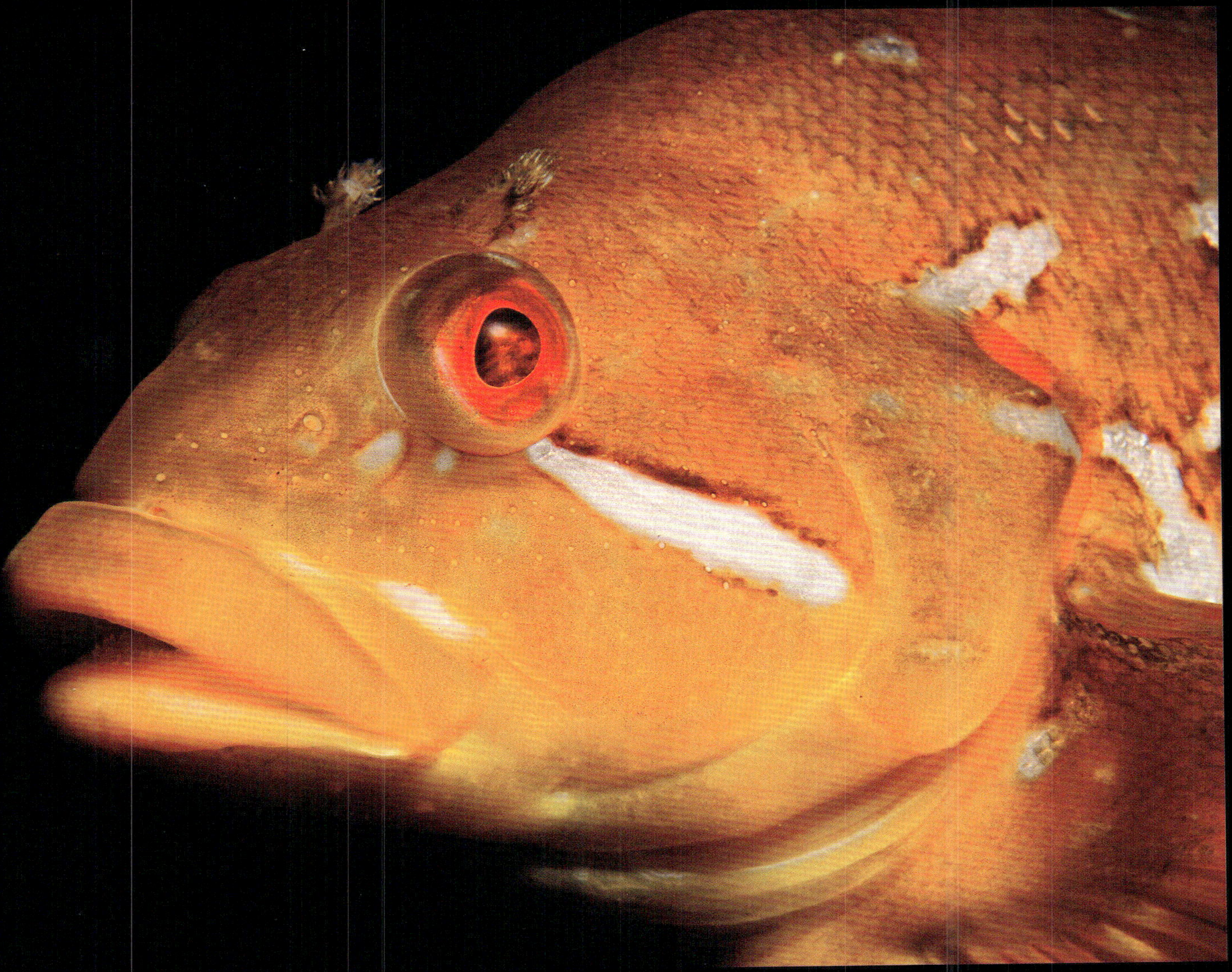

The rock greenling (*Hexagrammos lagocephalus*) is a master at blending in with its shallow, exposed surroundings. Although rock greenlings are common throughout the North Pacific, divers do not see them often because of the fish's camouflage ability and shy behavior. This species exhibits a distinct sexual dichromatism (different colors). The adult female is olive green, golden, or orange and the adult male is a brilliant array of colors—black, orange, white, and blue, with red blotches on sides. Mouths of adults are usually bluish inside. *Usof Bay, Unalaska Island, S. Harper photo.*

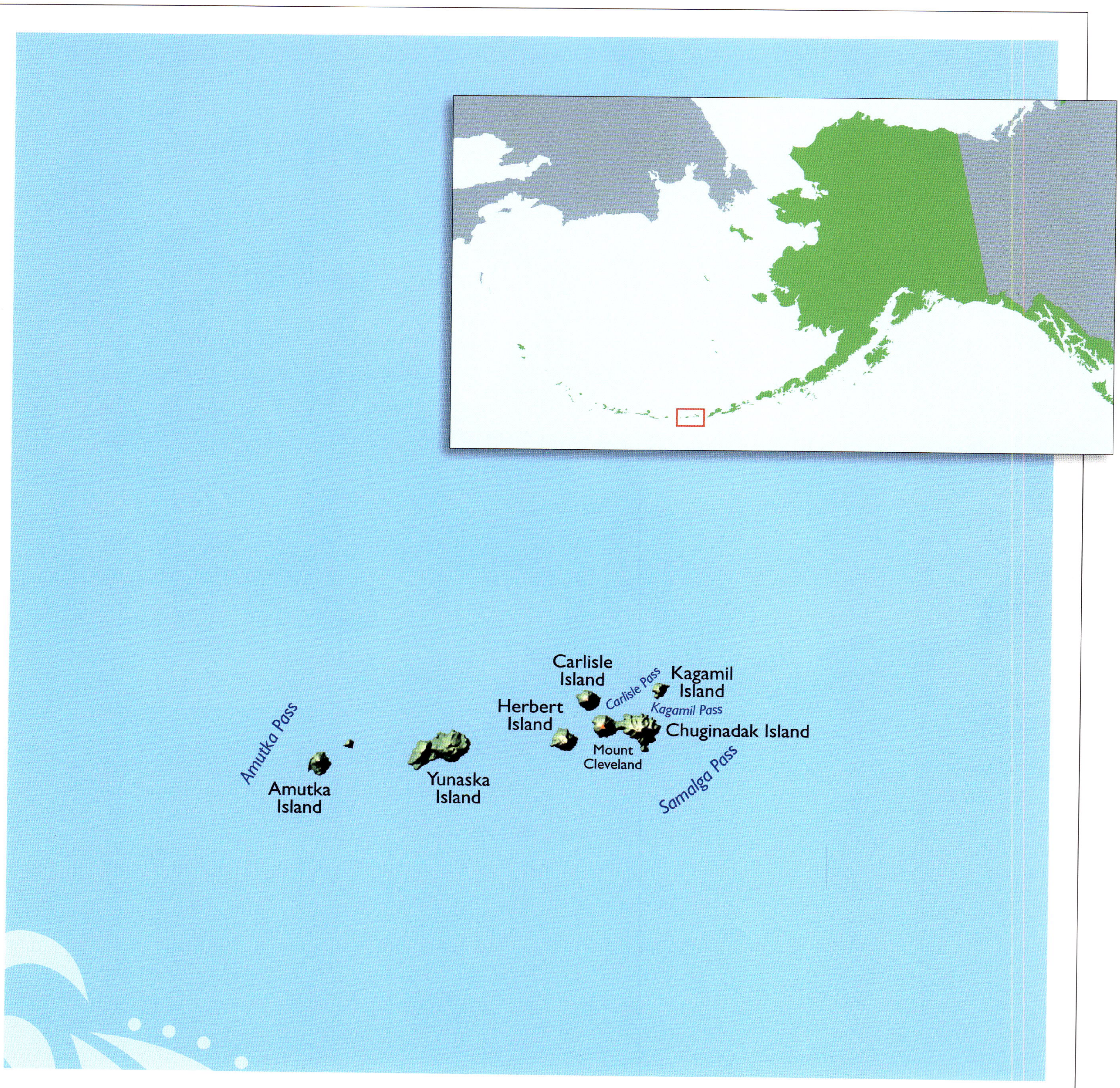
Carlisle Island
Carlisle Pass
Kagamil Island
Kagamil Pass
Herbert Island
Chuginadak Island
Amutka Pass
Mount Cleveland
Samalga Pass
Amutka Island
Yunaska Island

Islands of Four Mountains

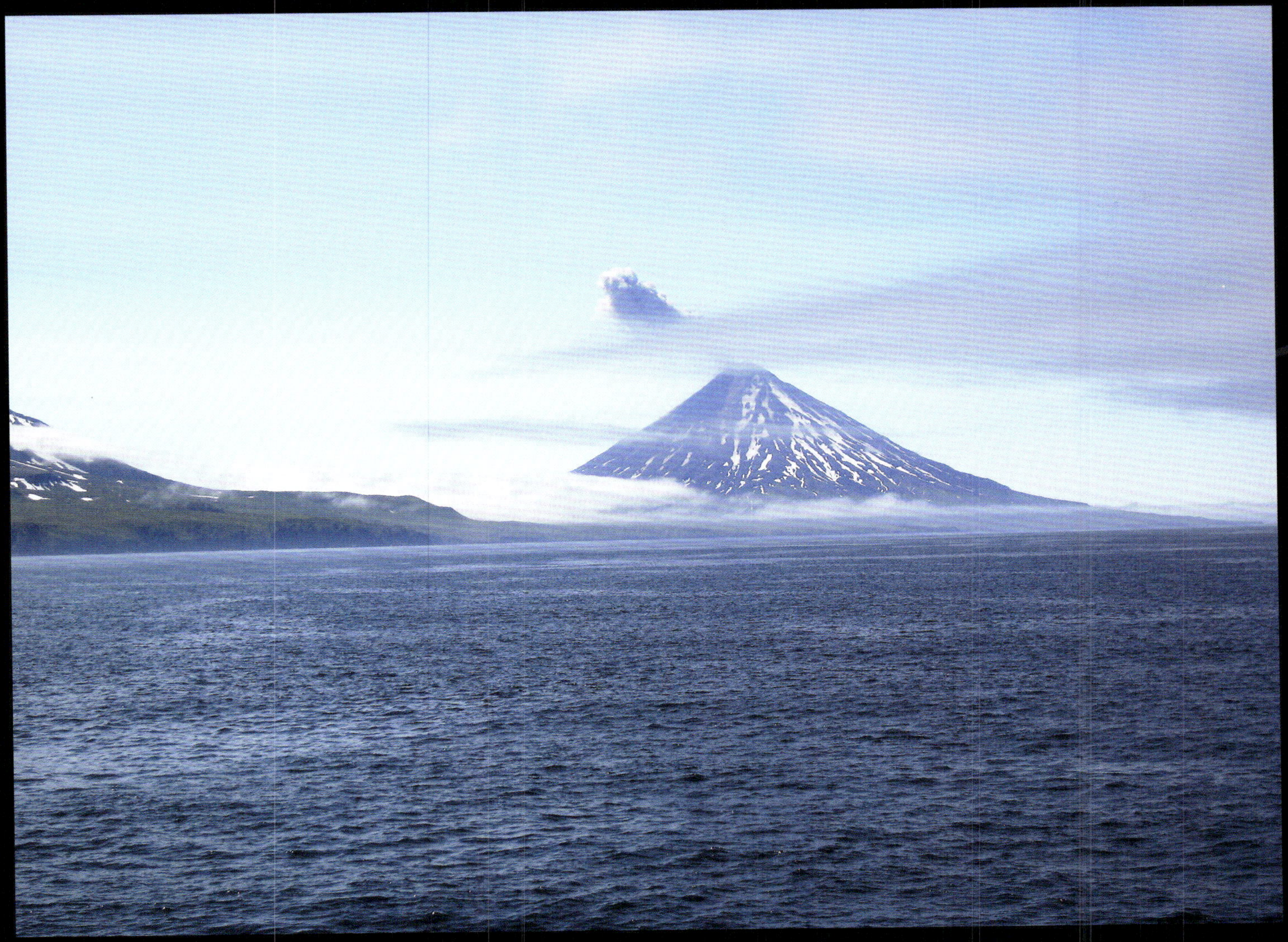

Mount Cleveland volcano on Chuginadak Island periodically erupts, belching smoke and ash. This photo was taken on July 20, 2007. Mount Cleveland (5,676 ft or 1,730 m) is one of four active volcanoes in the Islands of Four Mountains. *S. Jewett photo.*

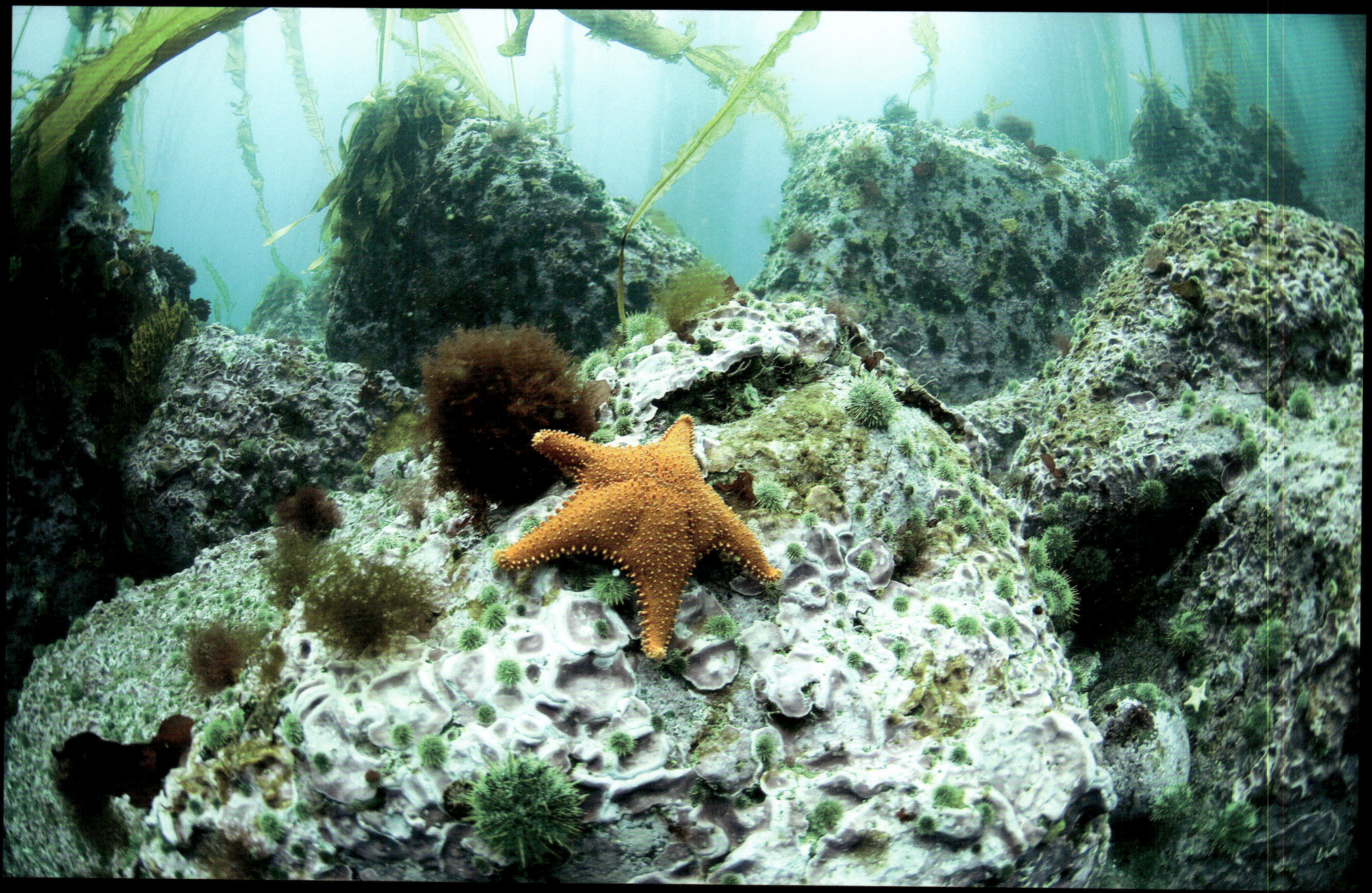

*H**ippasteria aleutica* sea stars are covered in spines, and their bony plates are cushion-like in appearance. These sea stars are usually reddish-orange in color with five short, thick arms. This newly named species is often found in exposed areas and presumably feeds on sea anemones, nudibranch eggs, sea slugs, and sea squirts, as does the similar *H. spinosa. Samalga Pass, Chuginadak Island, S. Harper photo.*

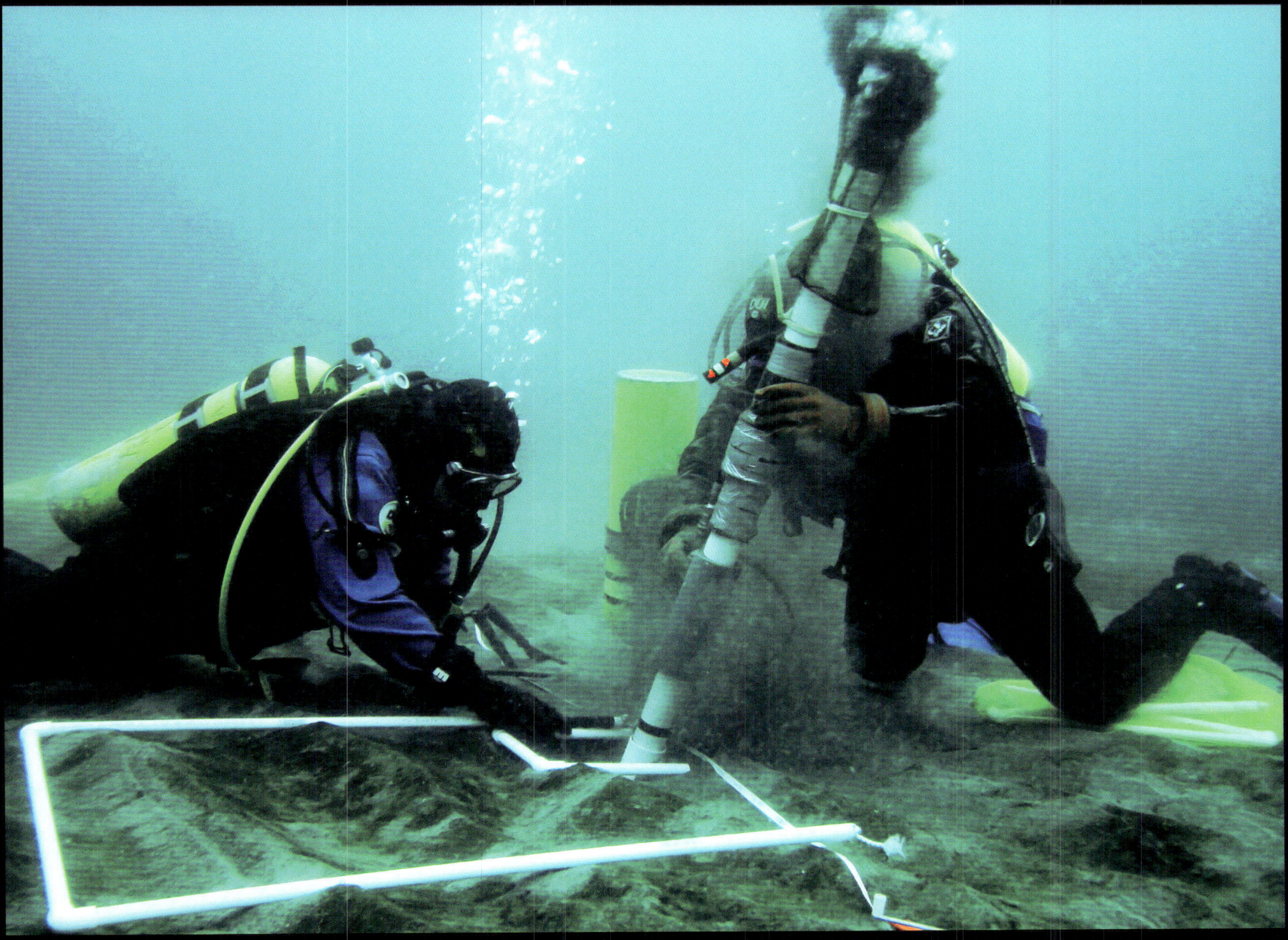

Divers sample sediment for infaunal invertebrates that live in mud and sand. Divers are using a custom-made suction dredge that vacuums sediment from the seafloor and is filtered by a catch bag at the other end. *Carlisle Pass, Carlisle Island, H. Chenelot photo.*

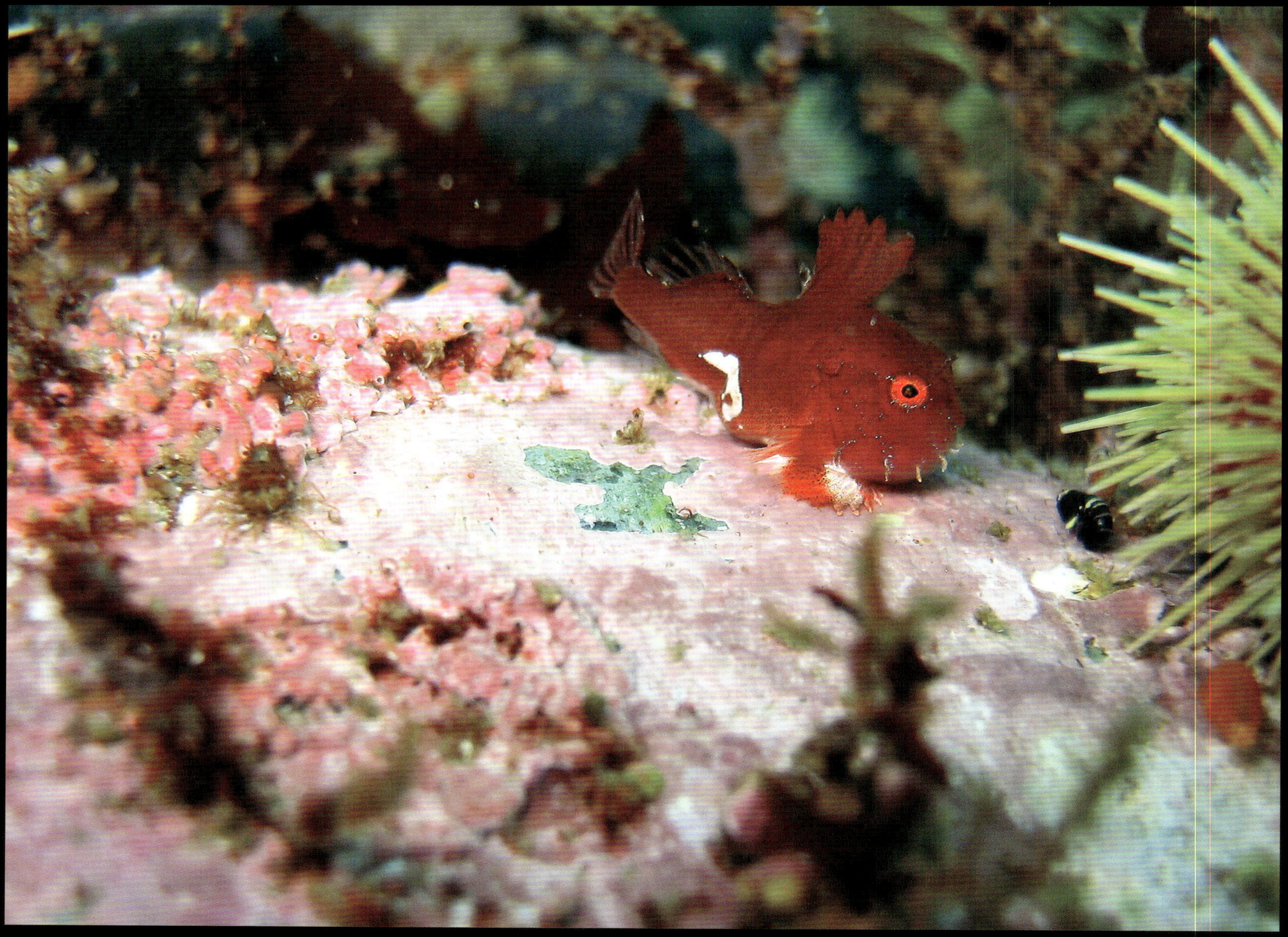

The Alaskan lumpsucker (*Eumicrotremus gyrinops*) has modified pelvic fins that form a large suction disk on its stomach, which it uses to attach to the substrate. This fish is unique because of its red color and small size (to 2 in, 5 cm total length). Lumpsuckers are found in the cold waters of the Arctic, North Atlantic, and North Pacific oceans. Little is known of the behavior or biology of the 27 identified species of lumpsuckers. *Kagamil Pass, Kagamil Island, H. Chenelot photo.*

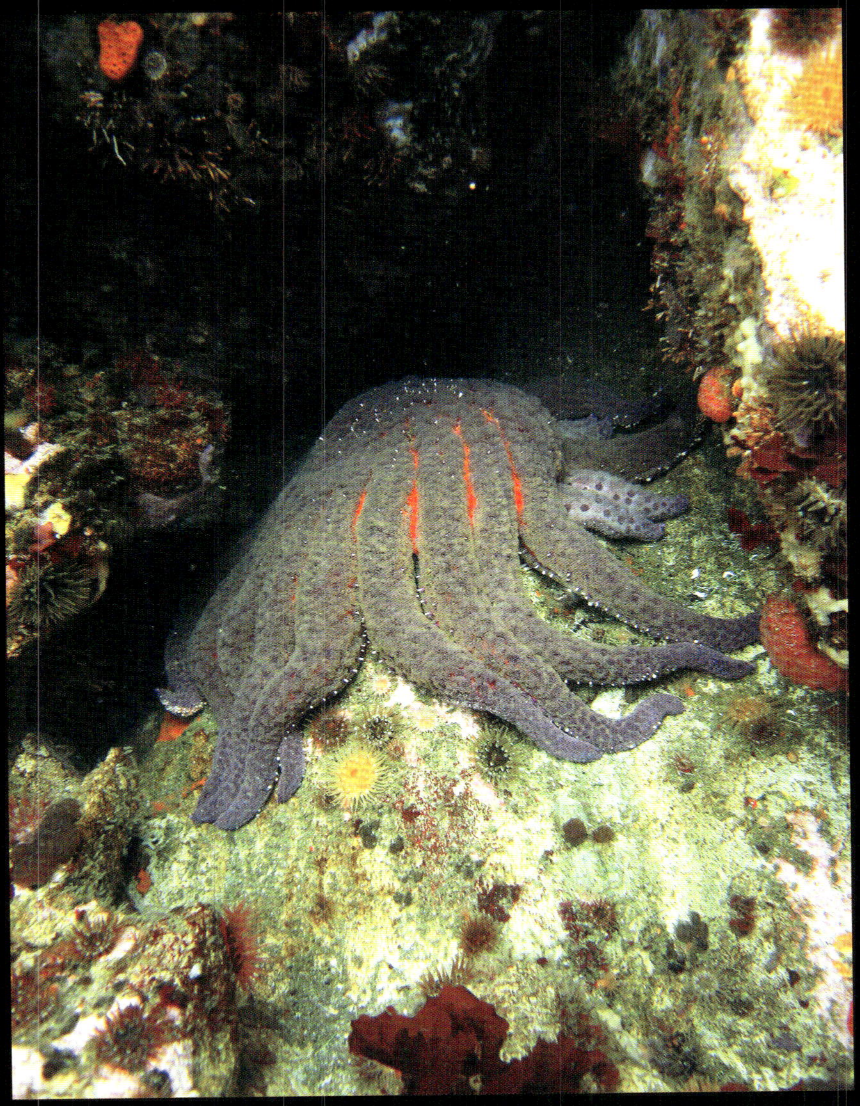

The sunflower star (*Pycnopodia helianthoides*) is one of the largest sea stars worldwide, sometimes larger than 3 ft (1 meter) across. It is also one of the fastest sea stars, moving to 30 in per minute (0.03 mph). The sunflower star usually has between 22 and 26 arms and more than 15,000 tubefeet. It is both a predator and a scavenger and is found from the central Aleutians to northern Mexico. *Samalga Pass, Chuginadak Island, R. Brewer photo.*

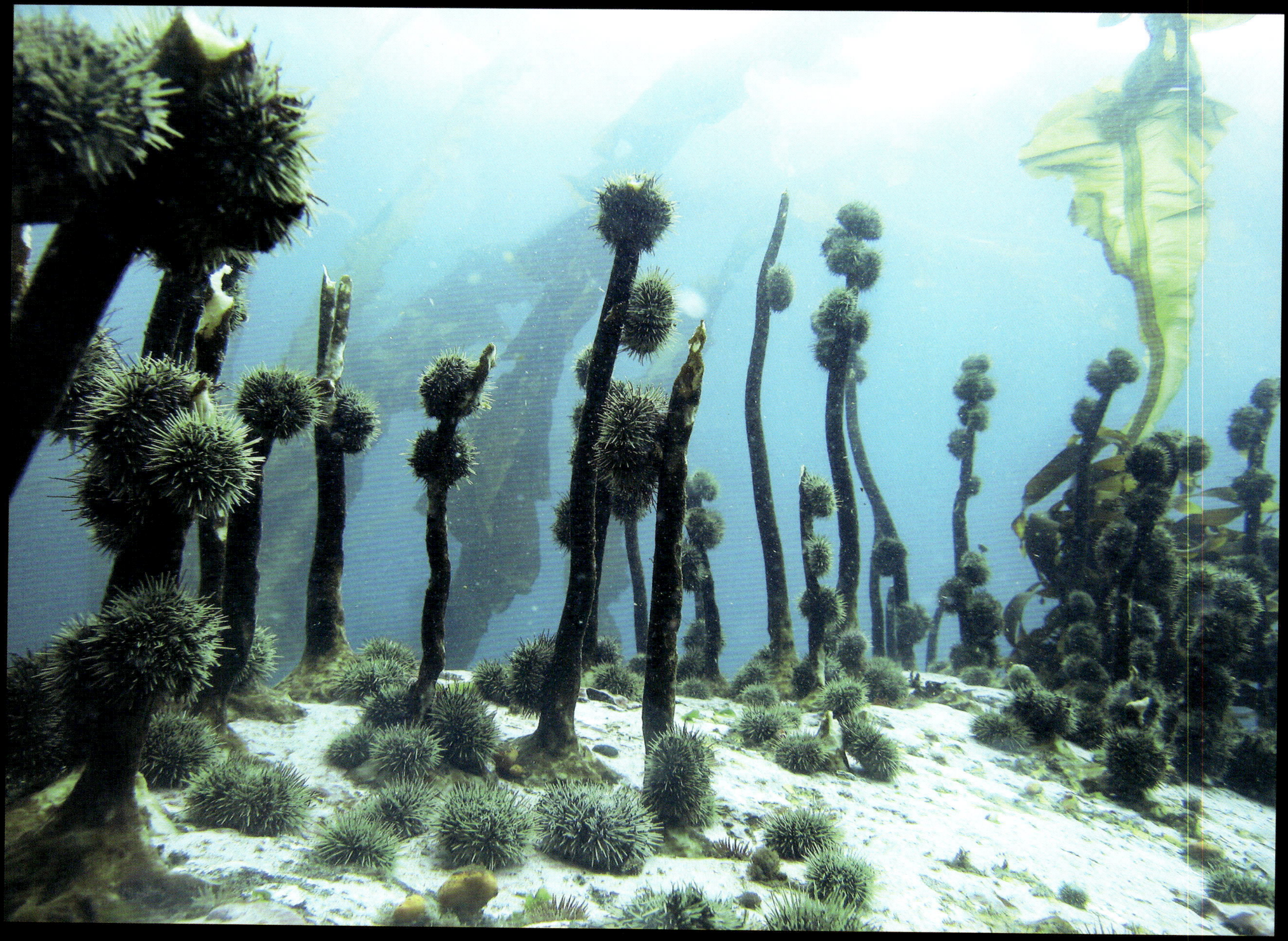

Green sea urchins (*Strongylocentrotus polyacanthus*) graze down a kelp forest of *Eualaria fistulosa.* The green sea urchin is a dominant herbivore and scavenger in the Aleutian nearshore ecosystem. Occasionally urchins become so numerous that they consume all the kelps in an area, excluding most other organisms and creating an "urchin barren." *Kagamil Pass, Kagamil Island, R. Brewer photo.*

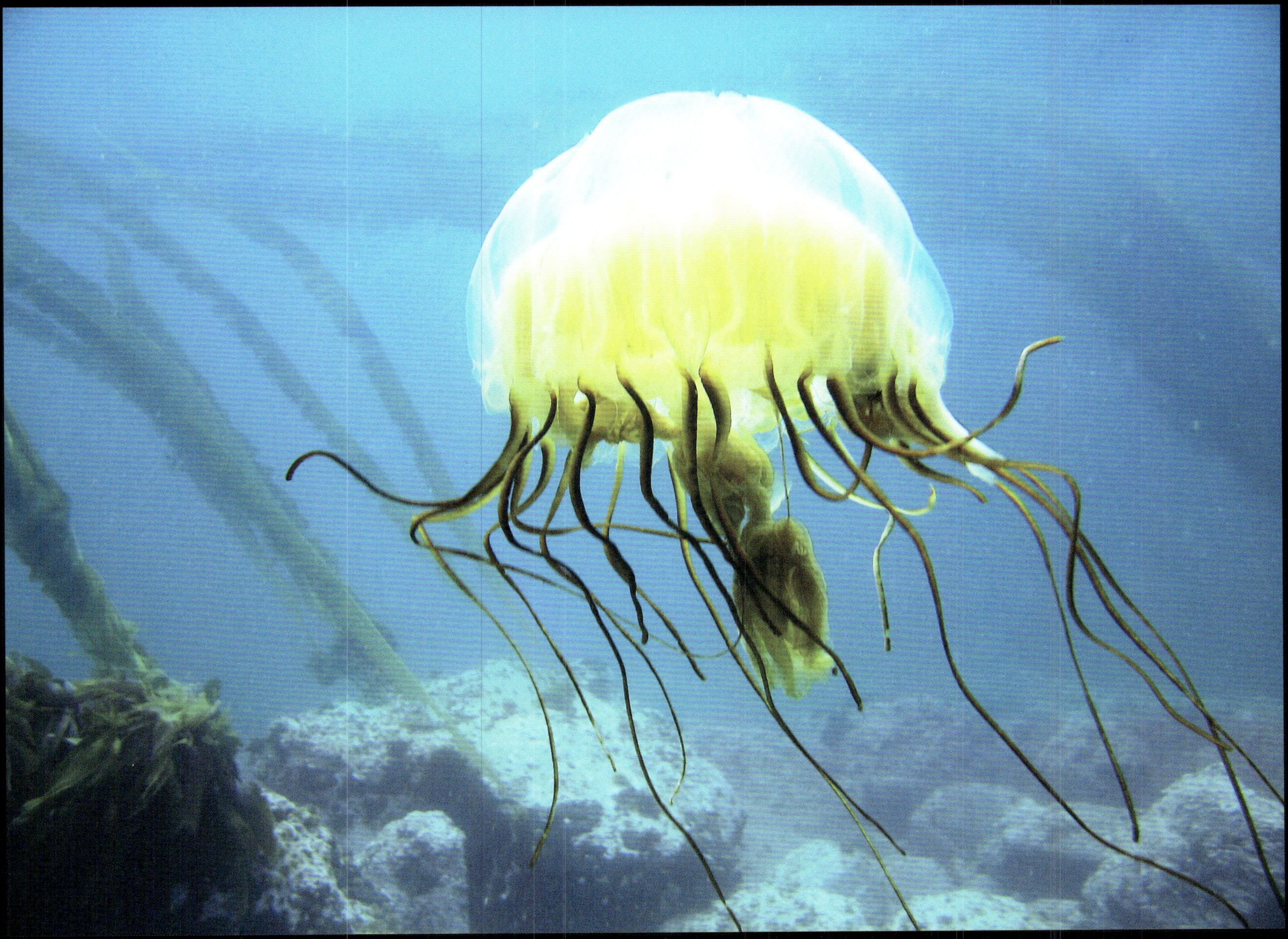

The sea nettle jellyfish (*Chrysaora melanaster*) is a common inhabitant of the open ocean of the Aleutian Islands. The bells of these gelatinous predators can reach 12 in (30 cm) across, with tentacles to 8 ft (2.4 m) long. *Kagamil Pass, Kagamil Island, S. Jewett photo.*

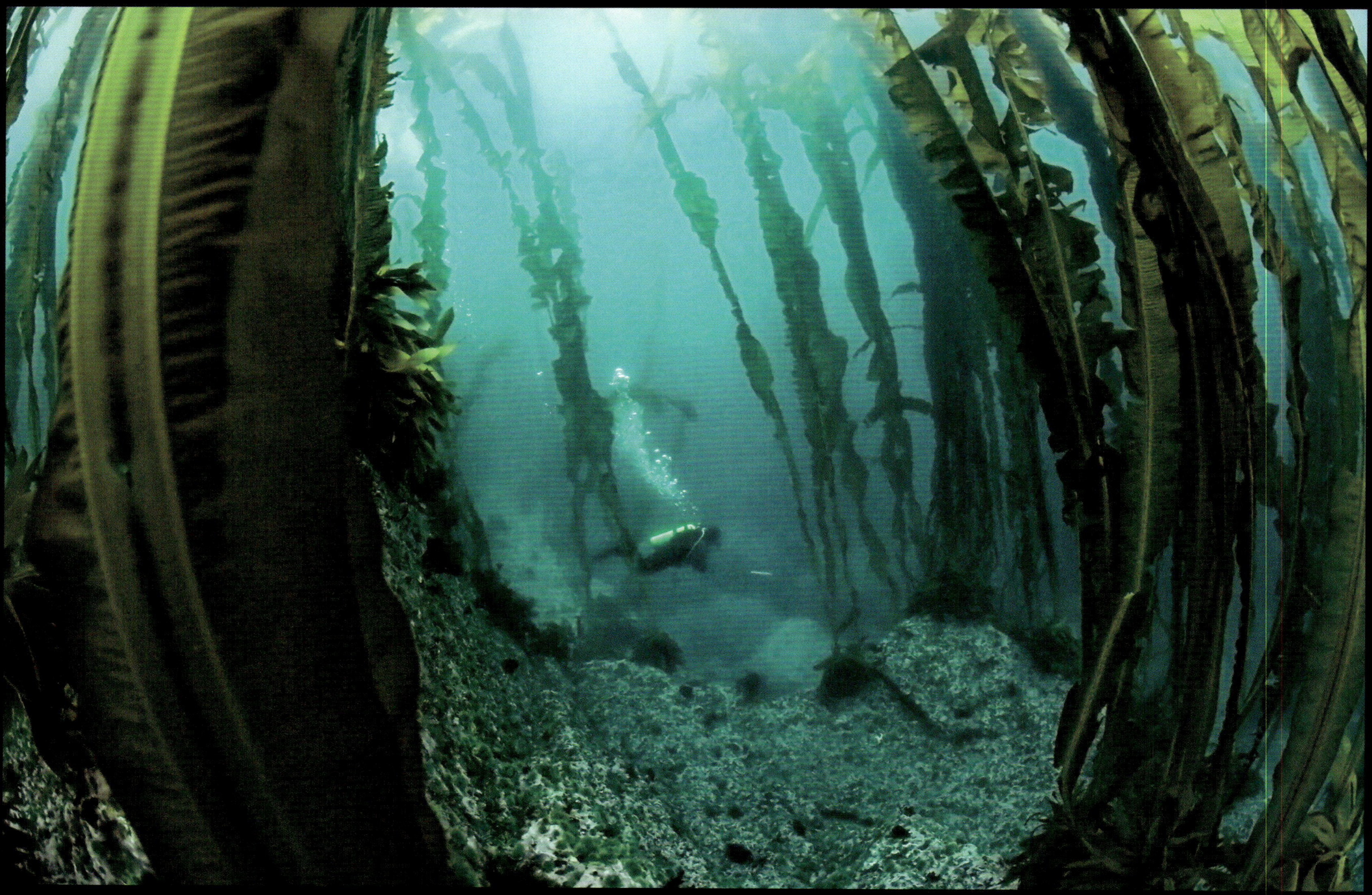

Although the Aleutian Islands are not known as scuba diving getaways, underwater vistas like this kelp forest of *Eualaria fistulosa* are common and worth the adventure to view. *Samalga Pass, Chuginadak Island, S. Harper photo.*

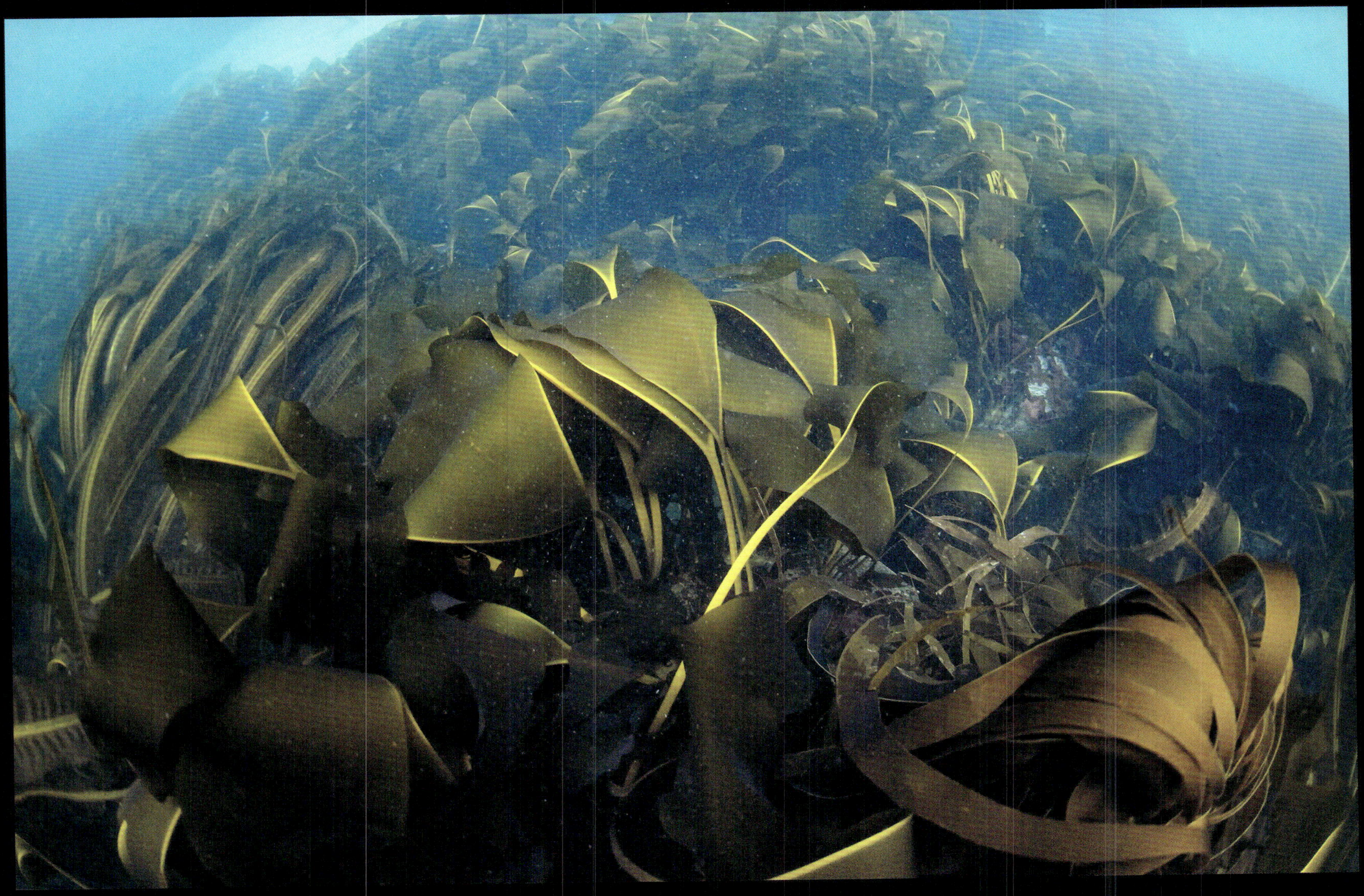

The golden V kelp (*Aureophycus aleuticus*) is a recently named new species discovered during the 2004-2007 expeditions. Researchers believe the golden V kelp may be an important link in revealing the evolution of the Alariaceae and Laminariaceae kelp families. *Kagamil Pass, Kagamil Island, S. Harper photo.*

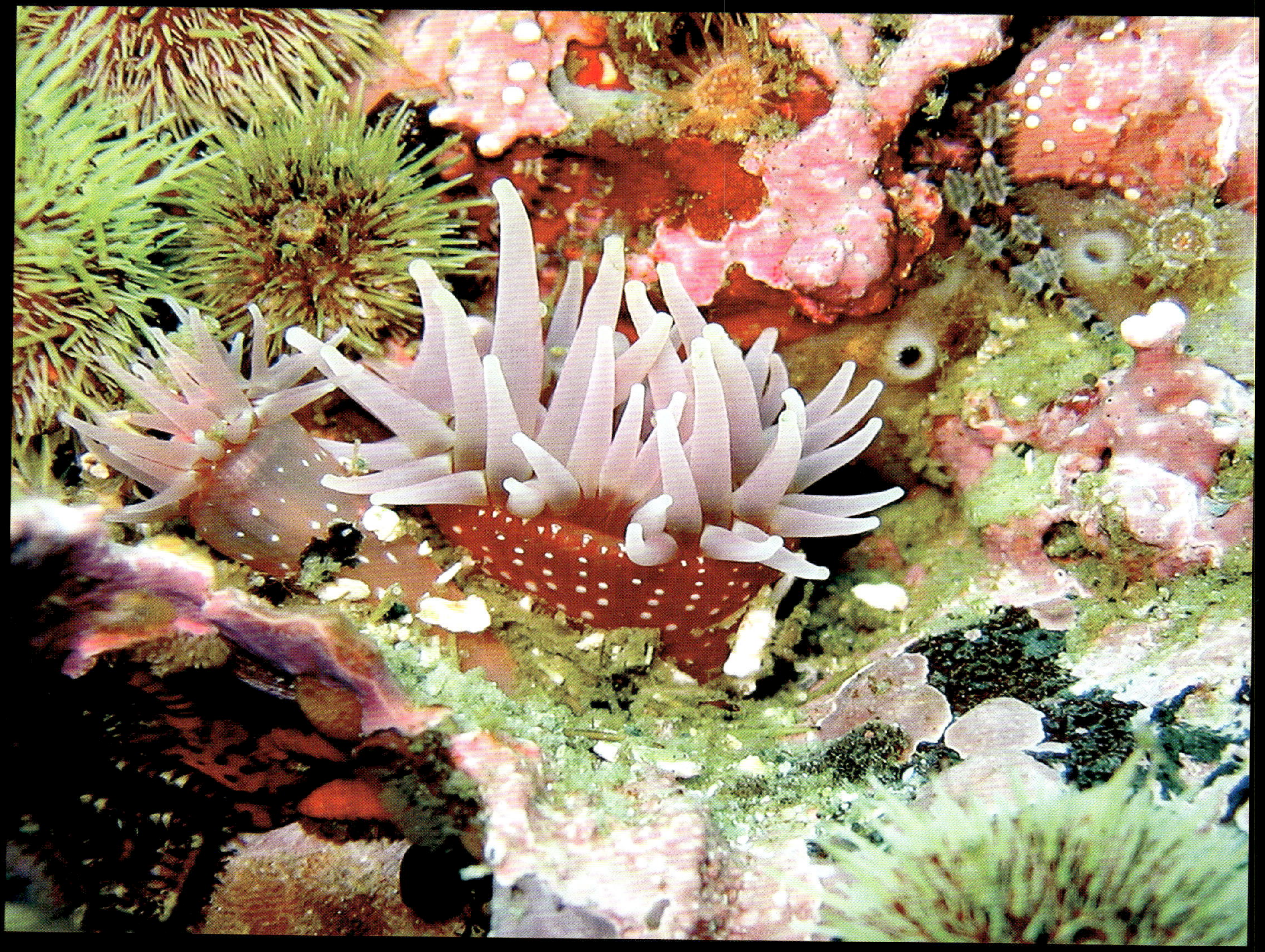

The white spotted rose anemone (*Urticina lofotensis*) has a distinctive bright pink column with irregular white spotting. This species is found attached to rocks, pilings, and marina floats; in the photo it hides in encrusting coralline algae. Sea anemones use stinging cells (nematocysts) located in their tentacles to paralyze prey and defend against predators. *Kagamil Island, H. Chenelot photo.*

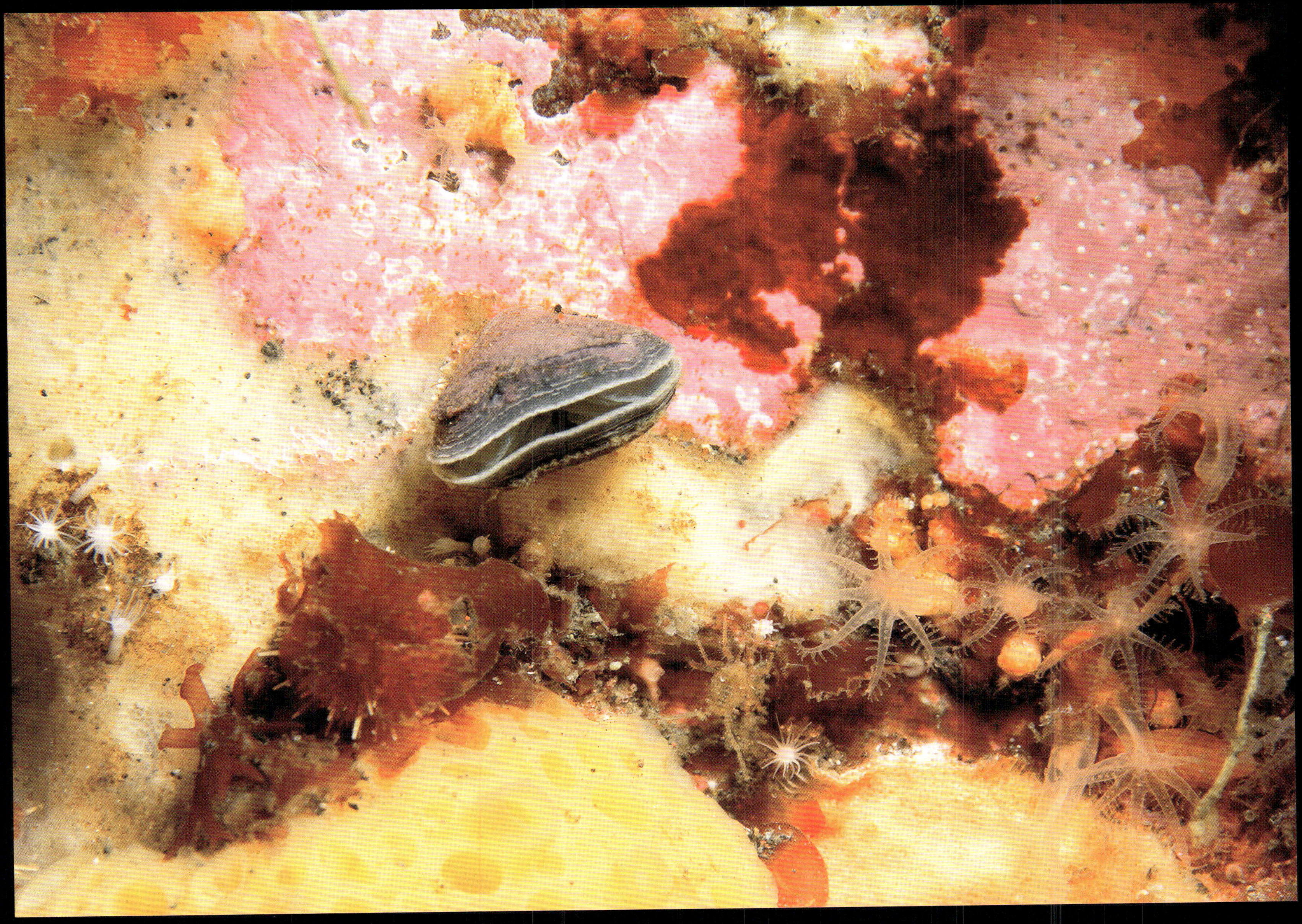

The black lamp shell (*Hemithyris psittacea*) looks similar to most bivalves, but is hinged on a stalk called a pedicel. This species is found throughout Alaska seas from 30 to 4,300 ft (10 to 1,300 m) deep. Lamp shells filter plankton from water using a twin spiraled lophophore. *Yunaska Island, S. Harper photo.*

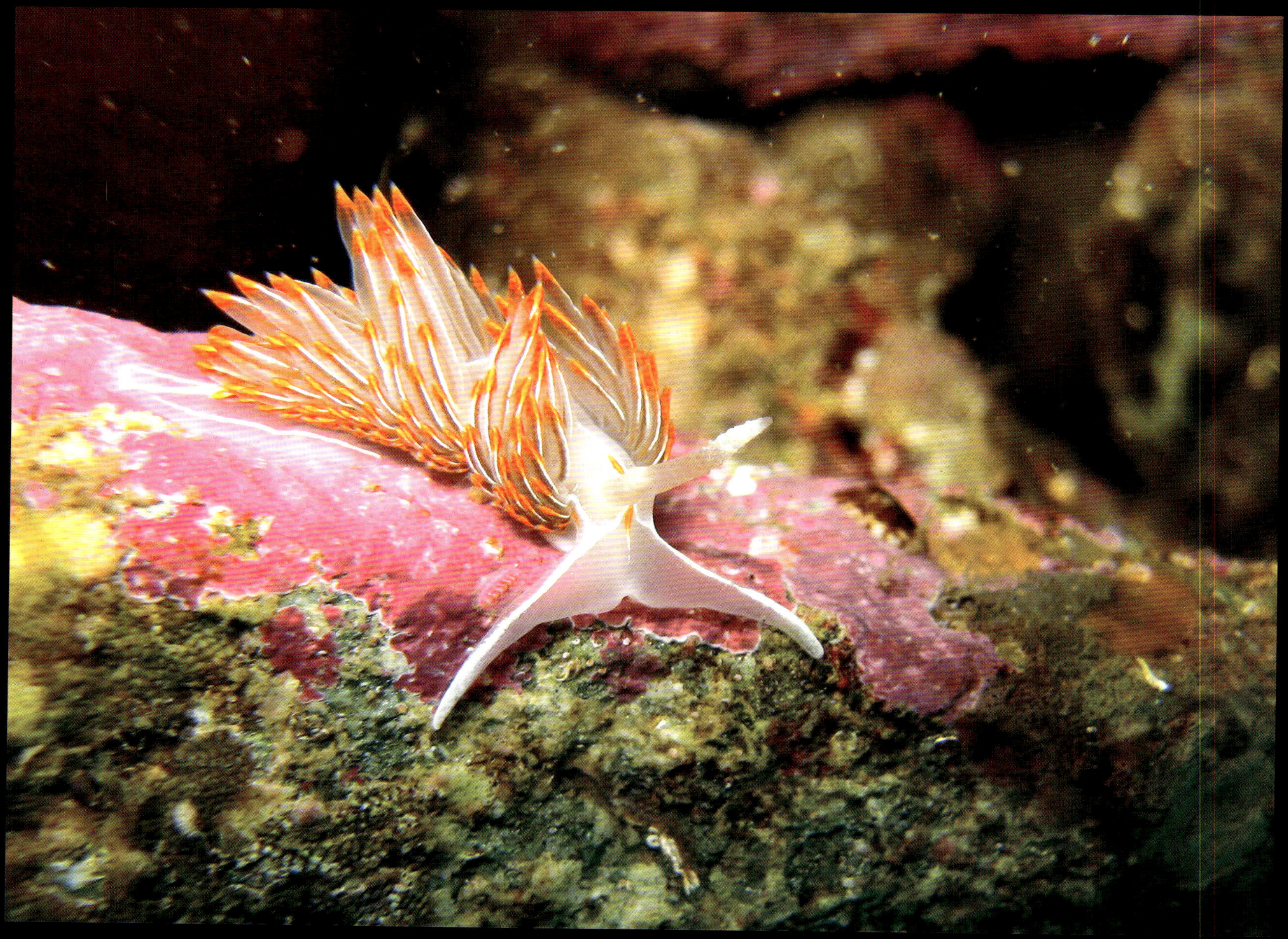

The opalescent nudibranch (*Hermissenda crassicornis*) has distinctive white lines on its head and orange-tipped cerata (gills). This species feeds mainly on hydroids and sea anemones. It incorporates undischarged nematocysts (stinging cells) from its prey into its cerata where they serve as a defense for the nudibranch. *Kagamil Pass, Kagamil Island, R. Brewer photo.*

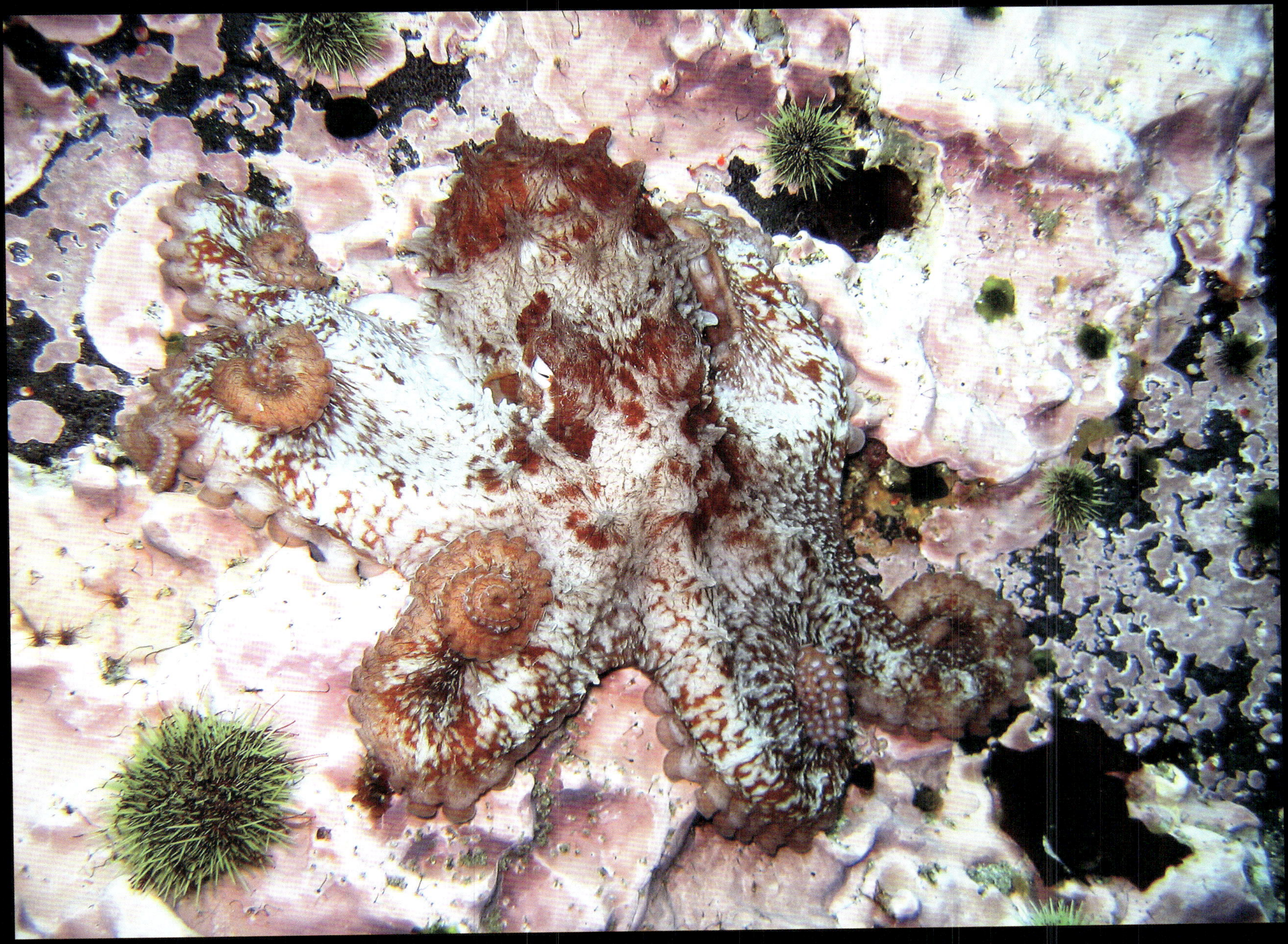

A North Pacific giant octopus (*Enteroctopus dofleini*) can grow to over 30 ft (10 m) in length and weigh up to 600 lbs (270 kg). This species is fast growing (up to 2% of its body weight per day) and lives for about five years. The male dies shortly after passing the female a sperm packet and the female dies after many weeks of tending her 20,000 to 50,000 eggs. The octopus in this photo has changed color to blend in with encrusting pink coralline alga. *Carlisle Island, H. Chenelot photo.*

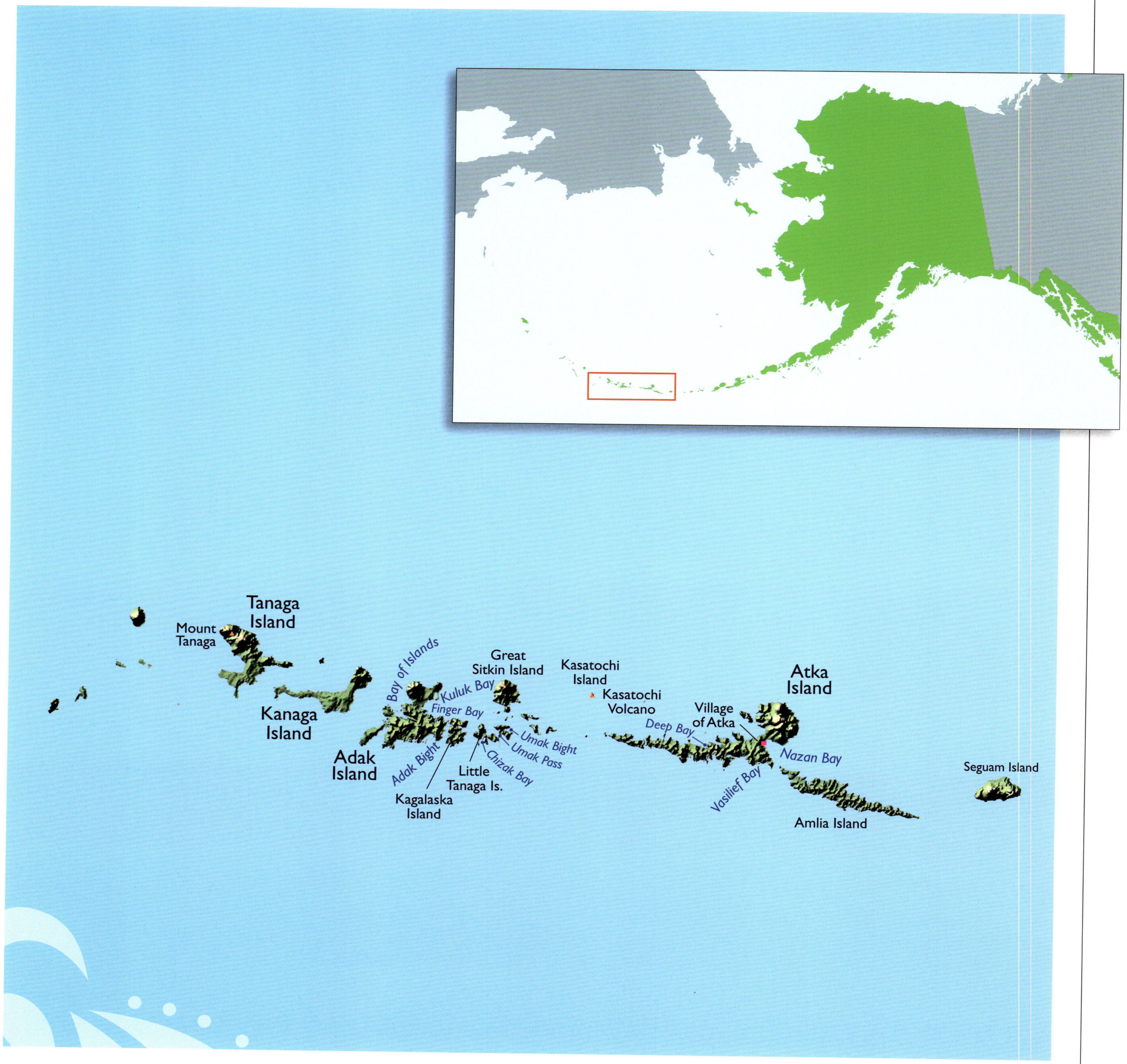
Tanaga Island
Mount Tanaga
Kanaga Island
Adak Island
Bay of Islands
Kuluk Bay
Finger Bay
Adak Bight
Kagalaska Island
Little Tanaga Is.
Chizak Bay
Umak Pass
Umak Bight
Great Sitkin Island
Kasatochi Island
Kasatochi Volcano
Atka Island
Village of Atka
Deep Bay
Nazan Bay
Vasilief Bay
Amlia Island
Seguam Island

Andreanof Islands

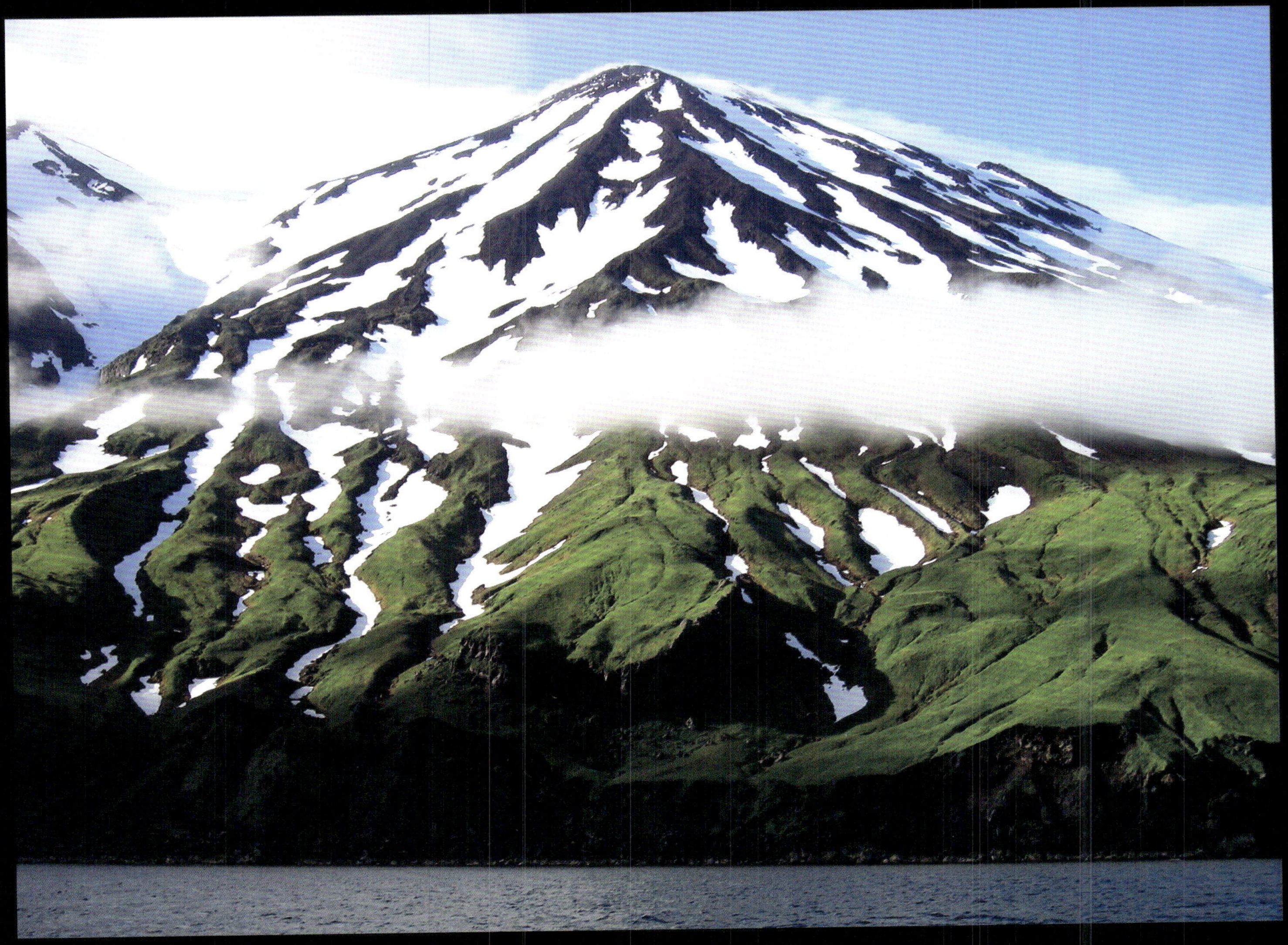

Tanaga Island has a land area of 204 mi^2 (528 km^2) making it the 33rd-largest island in the United States. The highest point on this uninhabited island is a volcano called Mount Tanaga, elevation 5,924 ft (1,806 m). *Tanaga Island, S. Jewett photo.*

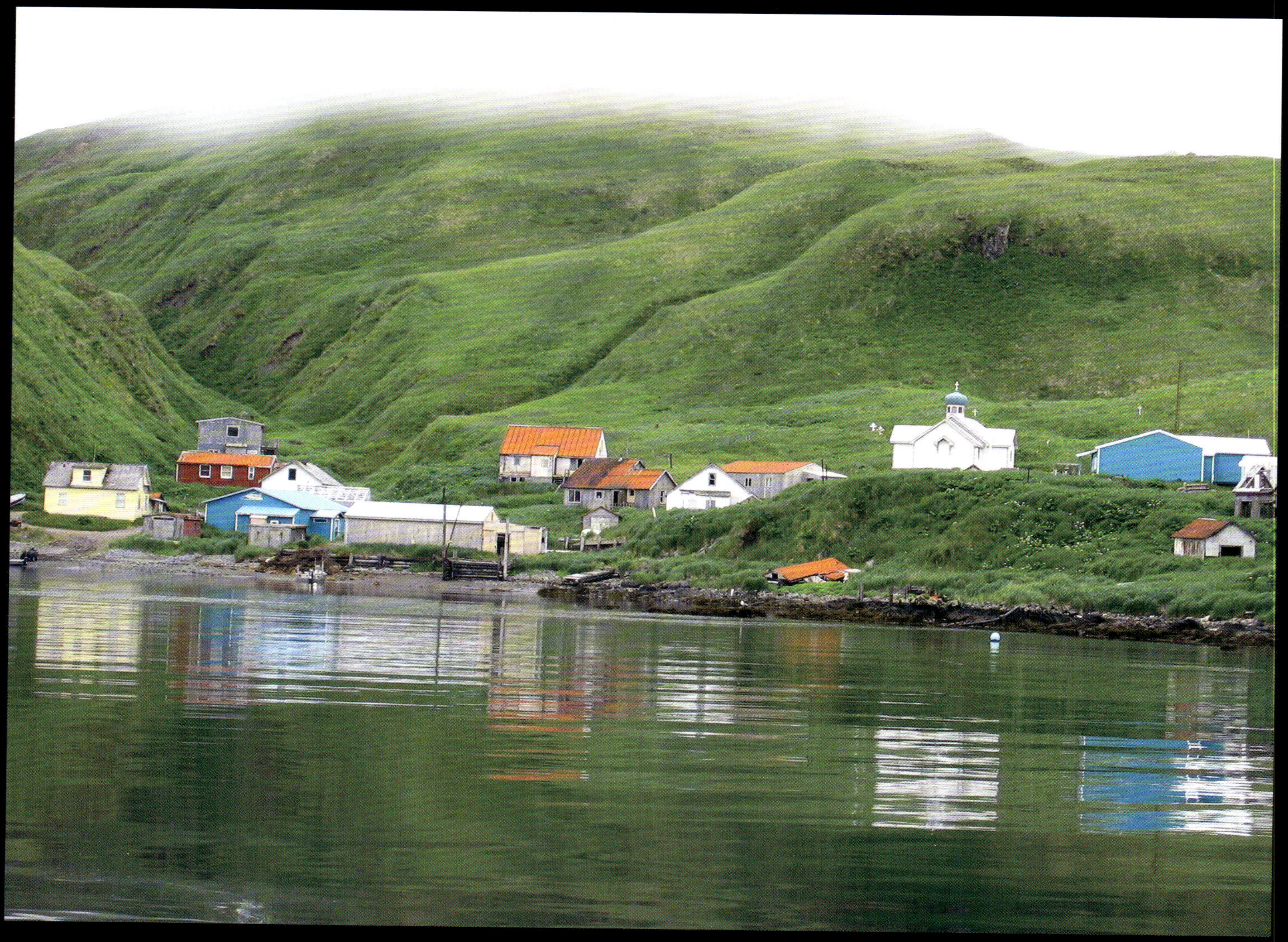

The village of Atka is located 1,200 air miles southwest of Anchorage, Alaska. Recent archaeological evidence suggests that the current site of the village has been occupied since prehistoric times. Atka Island has been inhabited by the Unangan people (Aleuts) for over 2,000 years, and now is home to fewer than 100 residents. *Village of Atka, Atka Island, S. Jewett photo.*

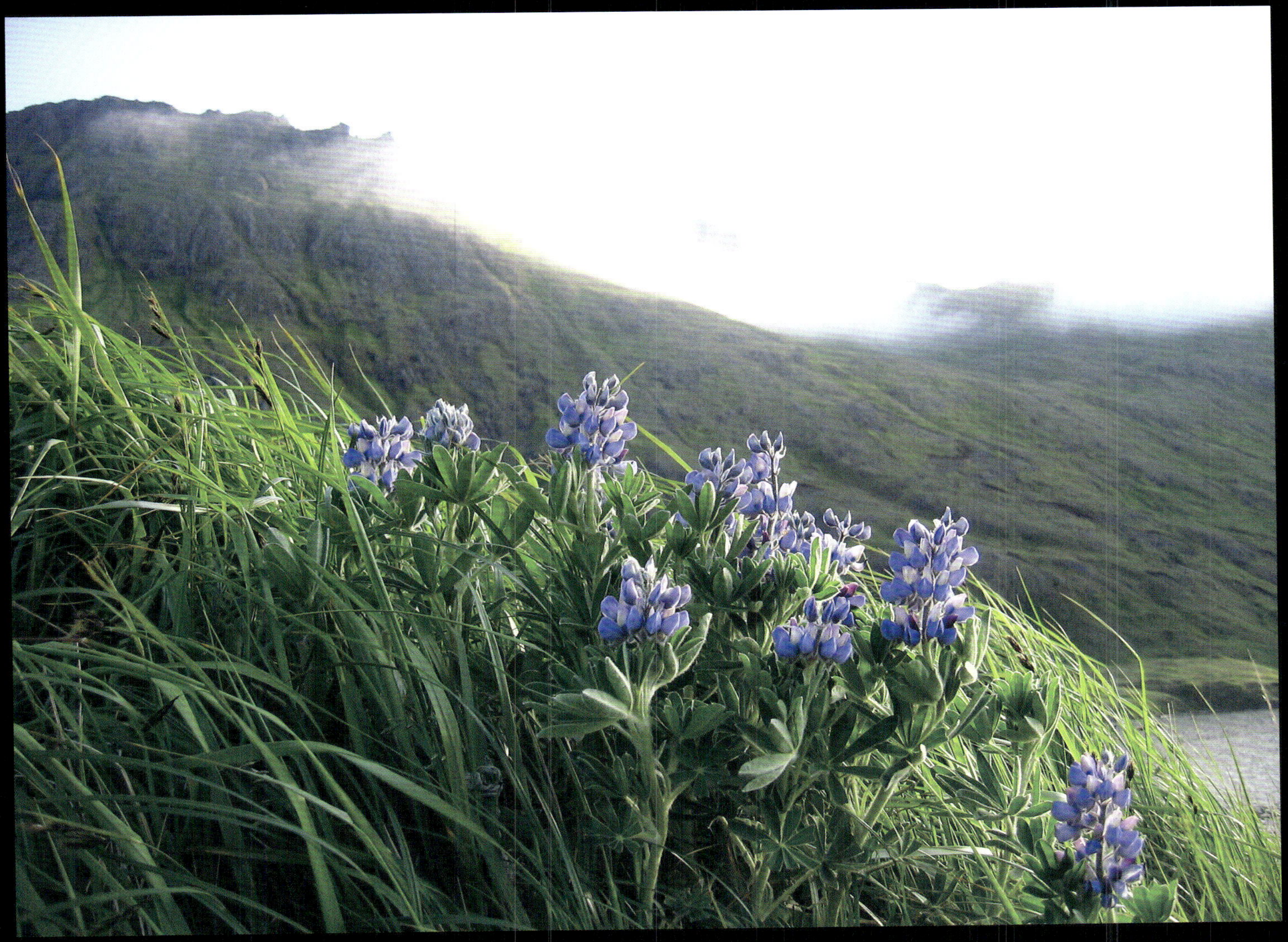

Arctic lupine (*Lupinus arcticus*) is a common sight throughout the Aleutian Islands in late summer. Lupines fix nitrogen from the atmosphere, which then fertilizes the soil for other plants. *Kagalaska Island, S. Jewett photo.*

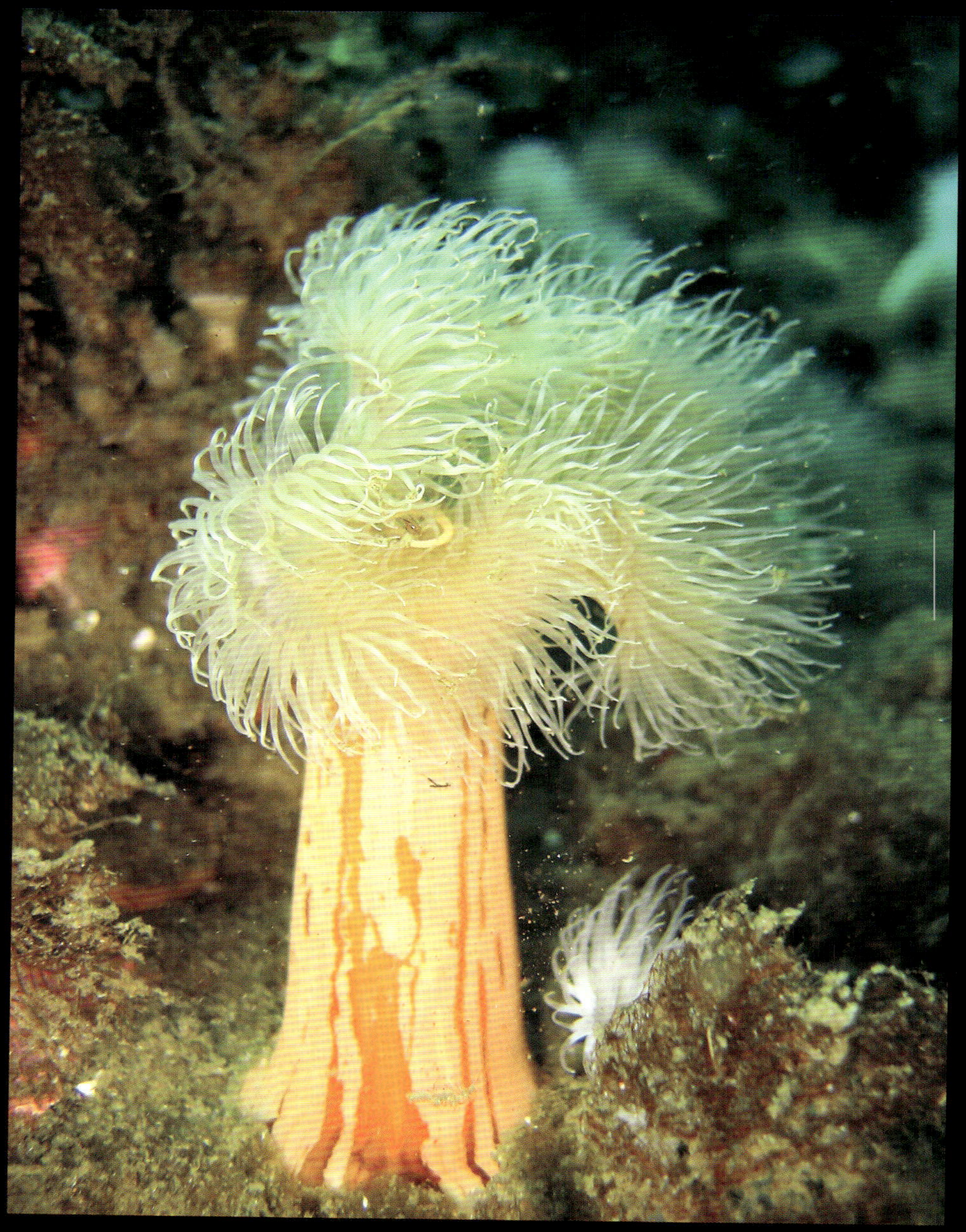

The small plumose anemone (*Metridium senile*) occurs along the eastern Pacific coast on rocks, shells, pilings, and docks where currents are strong. This species often reproduces asexually and is found in clusters. The primary food source for this species is zooplankton, though they also eat polychaetes, fish, and squid. *Kuluk Bay, Adak Island, H. Chenelot photo.*

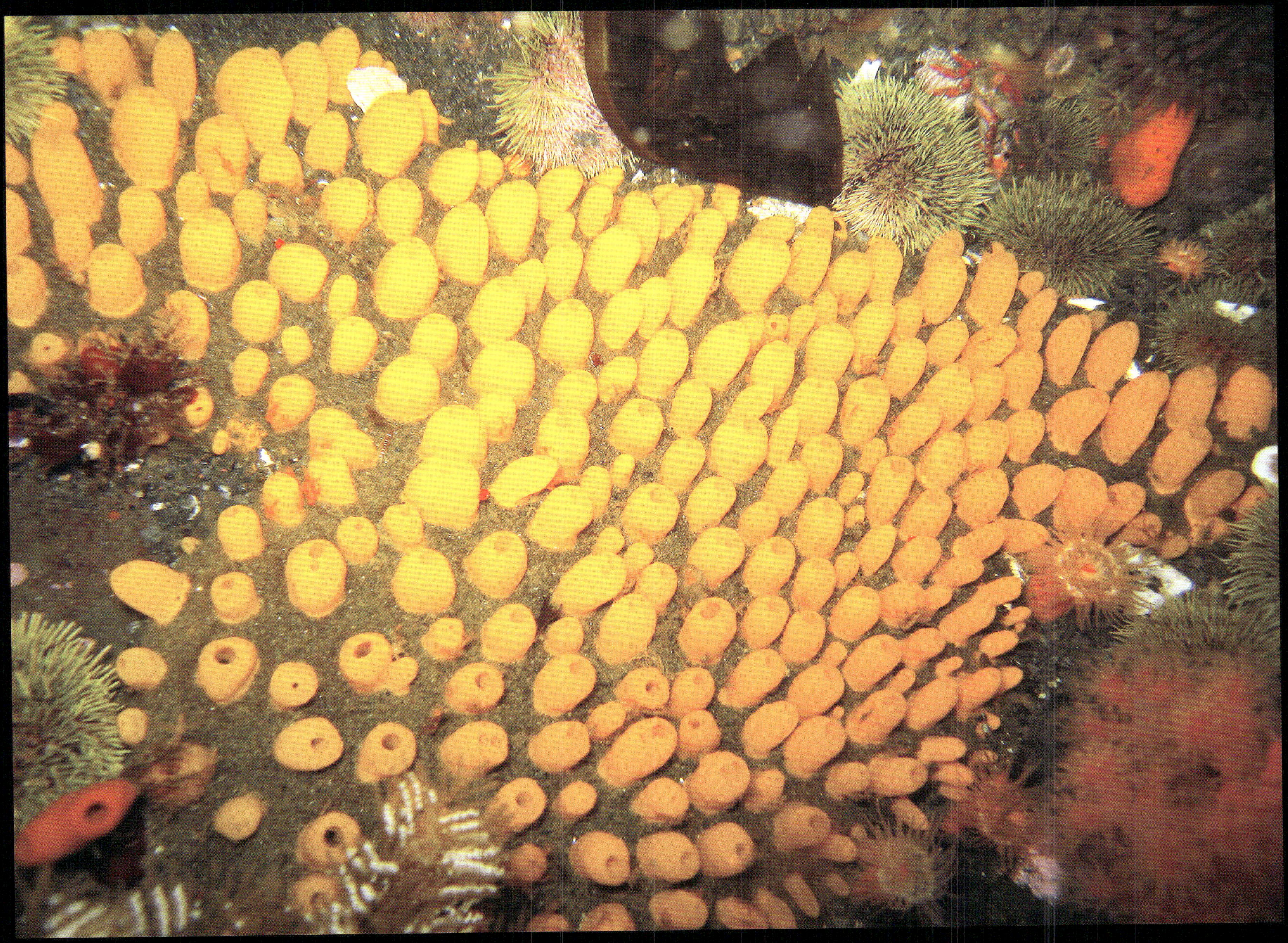

The aggregated nipple sponge (*Polymastia pacifica*) can cover an area up to 3 ft (1 m) across. The seemingly independent individuals are in fact one organism covered with a fine layer of sand. This species is found throughout the Aleutian Islands. *Tanaga Island, R. Brewer photo.*

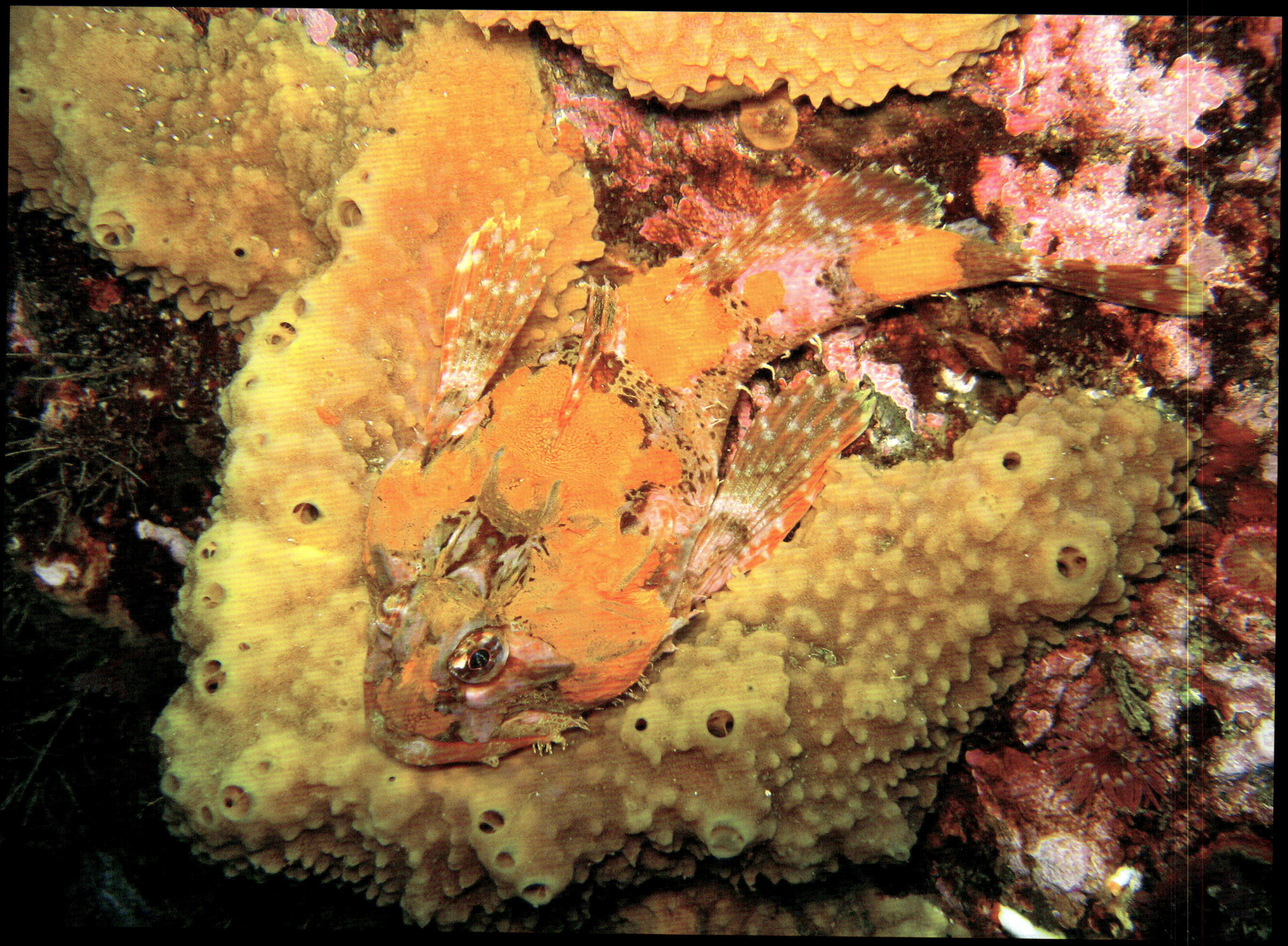

This antlered sculpin (*Enophrys diceraus*), with unusual pink and yellow patches, rests on a rock wall camouflaged with the surrounding sponge and coralline algae. This fish is distributed from the northern Chukchi Sea to southeast Alaska. *Deep Bay, Atka Island, S. Jewett photo.*

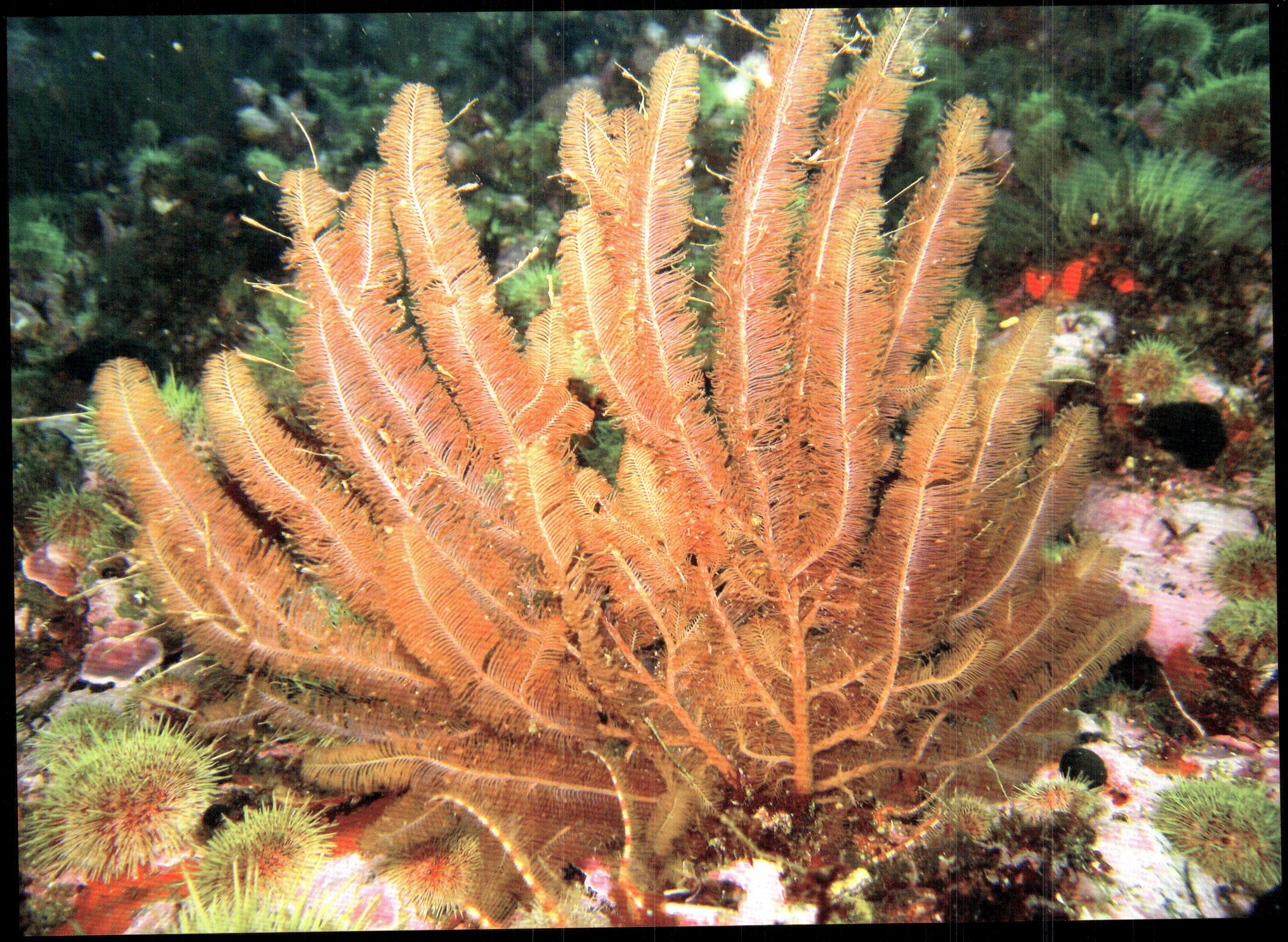

This unusually large (about 12 in x 12 in, 30 cm) colony of ostrich-plume hydroids (*Aglaophenia* sp.) is sometimes dismissed as seaweed, but these tiny predators are related to sea anemones, corals, and jellyfish. Each colony contains millions of minute polyps and is a microcosm for other small marine life, such as caprellid amphipods. *Bay of Islands, Adak Island, S. Jewett photo.*

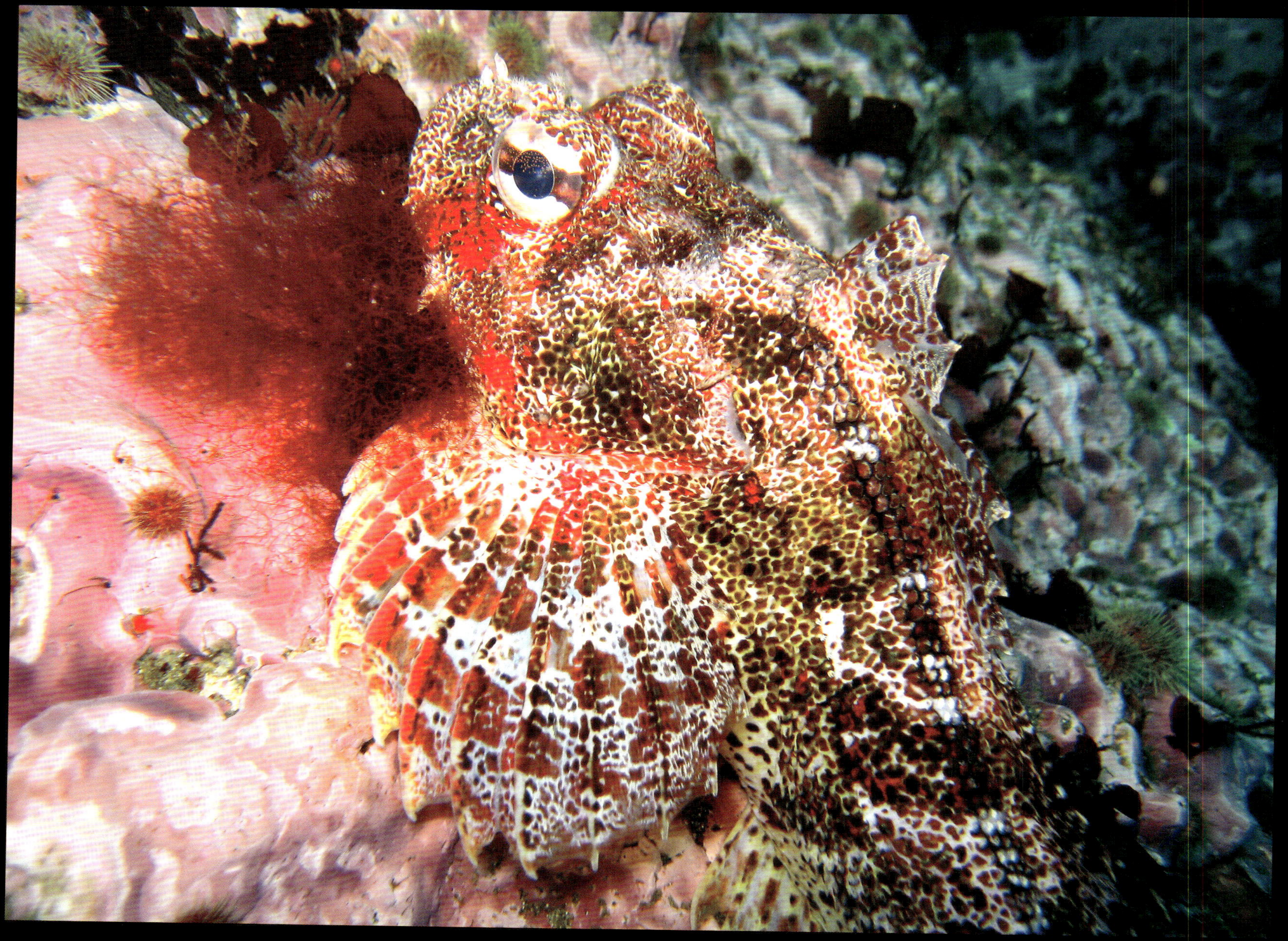

The red Irish lord (*Hemilepidotus hemilepidotus*) has only four or five rows of scales below the dorsal fin and several patches of red on the upper body. This species can grow to 20 in (50 cm) in length, and feeds on crabs, barnacles, and mussels. The red Irish lord is immobile in its camouflage state, and is often the subject of underwater photographs. *Nazan Bay, Atka Island, R. Brewer photo.*

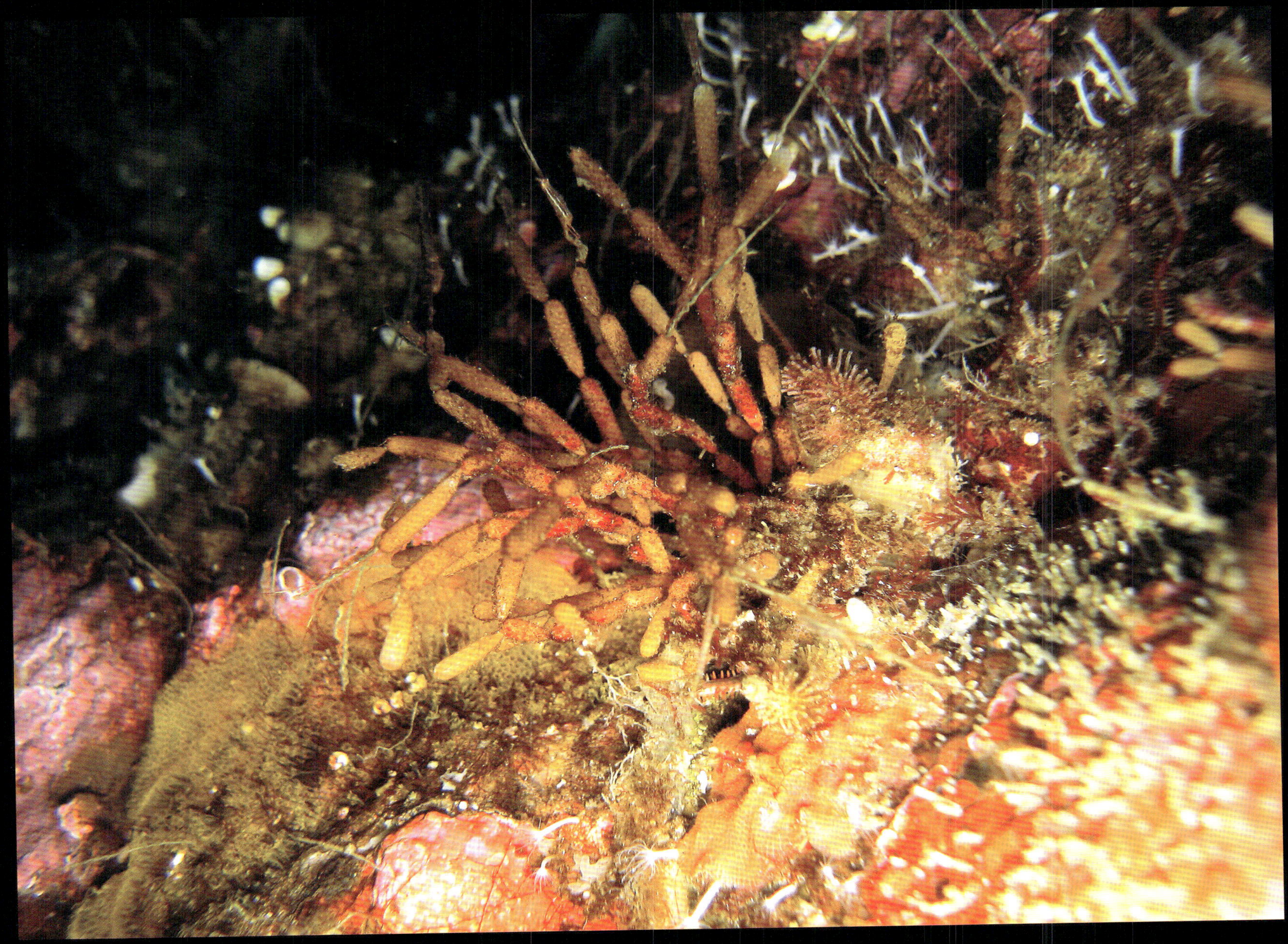

Stick bryozoans (*Microporina borealis*) are filter-feeders that have tiny feeding zooids along their entire length. This species is easily identified by the articulated appearance of each of the branches. Most bryozoan colonies like the one pictured here are hermaphroditic, possessing both male and female gonads. *Bay of Islands, Adak Island, R. Brewer photo.*

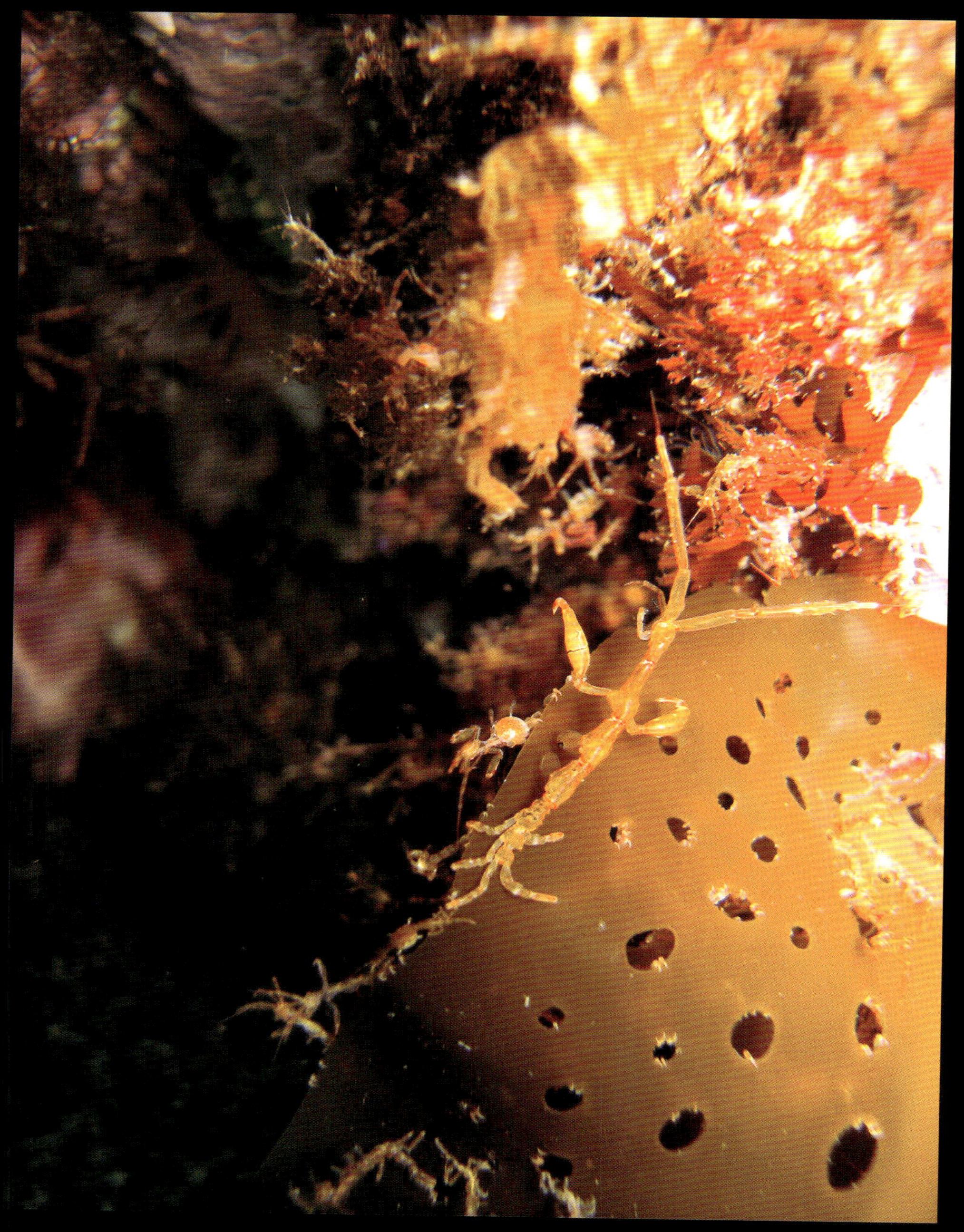

A skeleton shrimp (*Caprella* sp.) clings to a blade of sieve kelp (*Agarum* sp.) with ten grasping legs. It may be the largest skeleton shrimp in the Pacific Northwest, though it reaches only about 1.6 in (4 cm) in length. *Deep Bay, Atka Island, R. Brewer photo.*

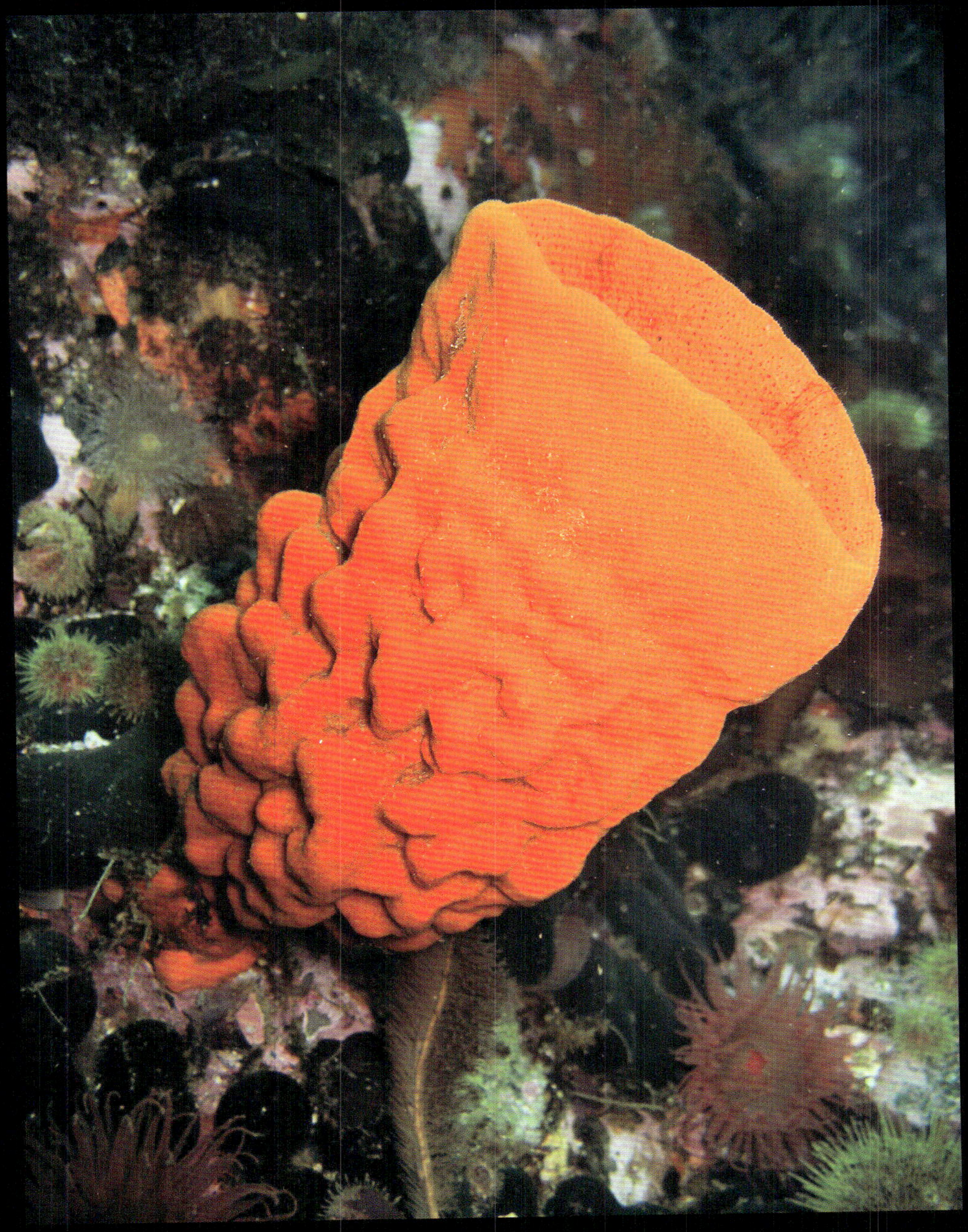

The orange funnel or vase sponge (*Stylissa* sp.) seems to prefer regions of high currents. Some individuals are 3 ft (1 m) high with a cone opening of 12 in (30 cm) or more. This sponge occurs throughout the Aleutian Islands. *Deep Bay, Atka Island, M. Hoberg photo.*

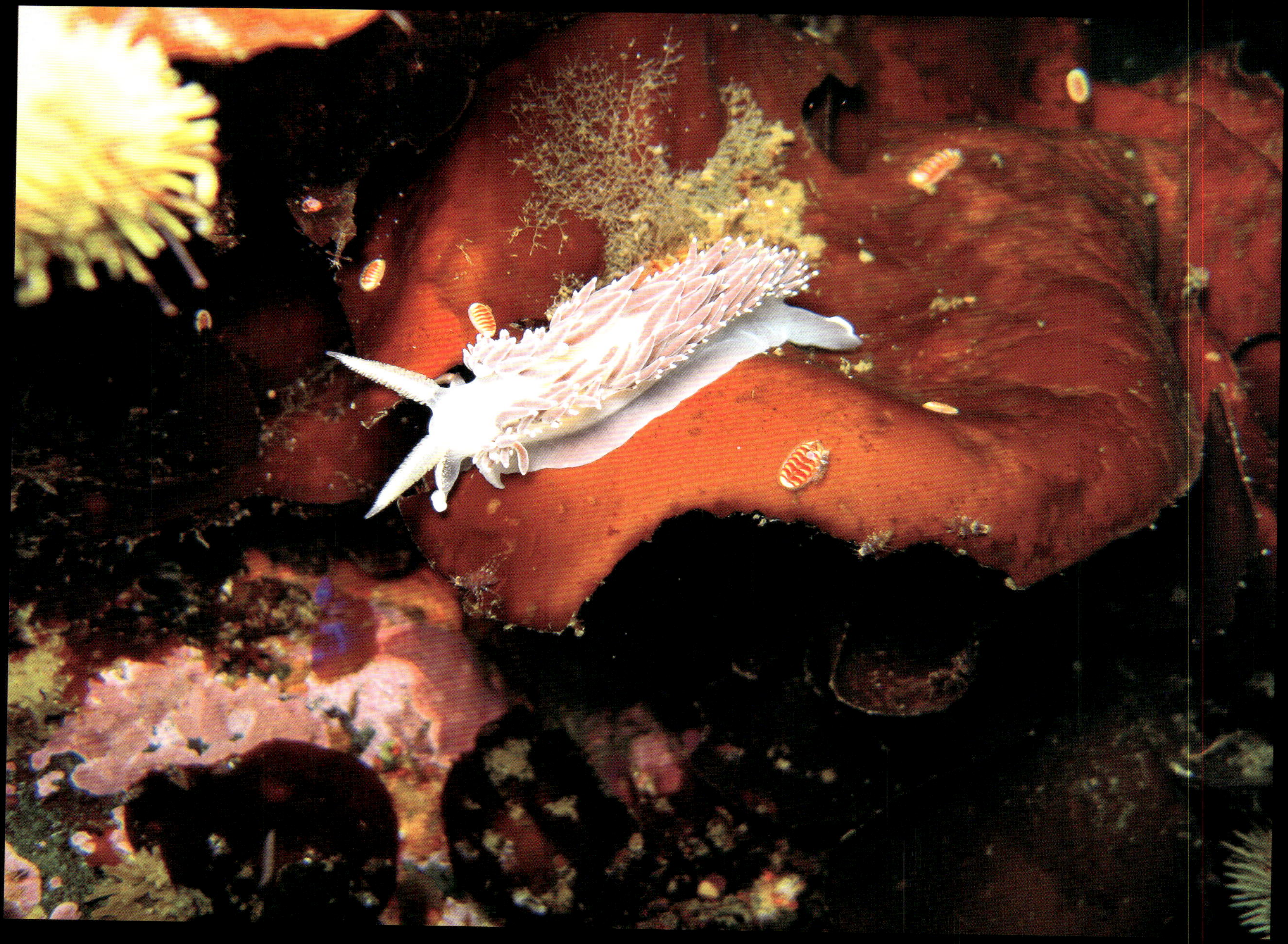

The red flabellina nudibranch (*Flabellina trophina*) has a distinctive pale red core and white tips on its gills. The two prominent rhinophores near its head are used for food and mate location. This species feeds mainly on hydroids. To swim it undulates its body back and forth to avoid predators. *Deep Bay, Atka Island, S. Jewett photo.*

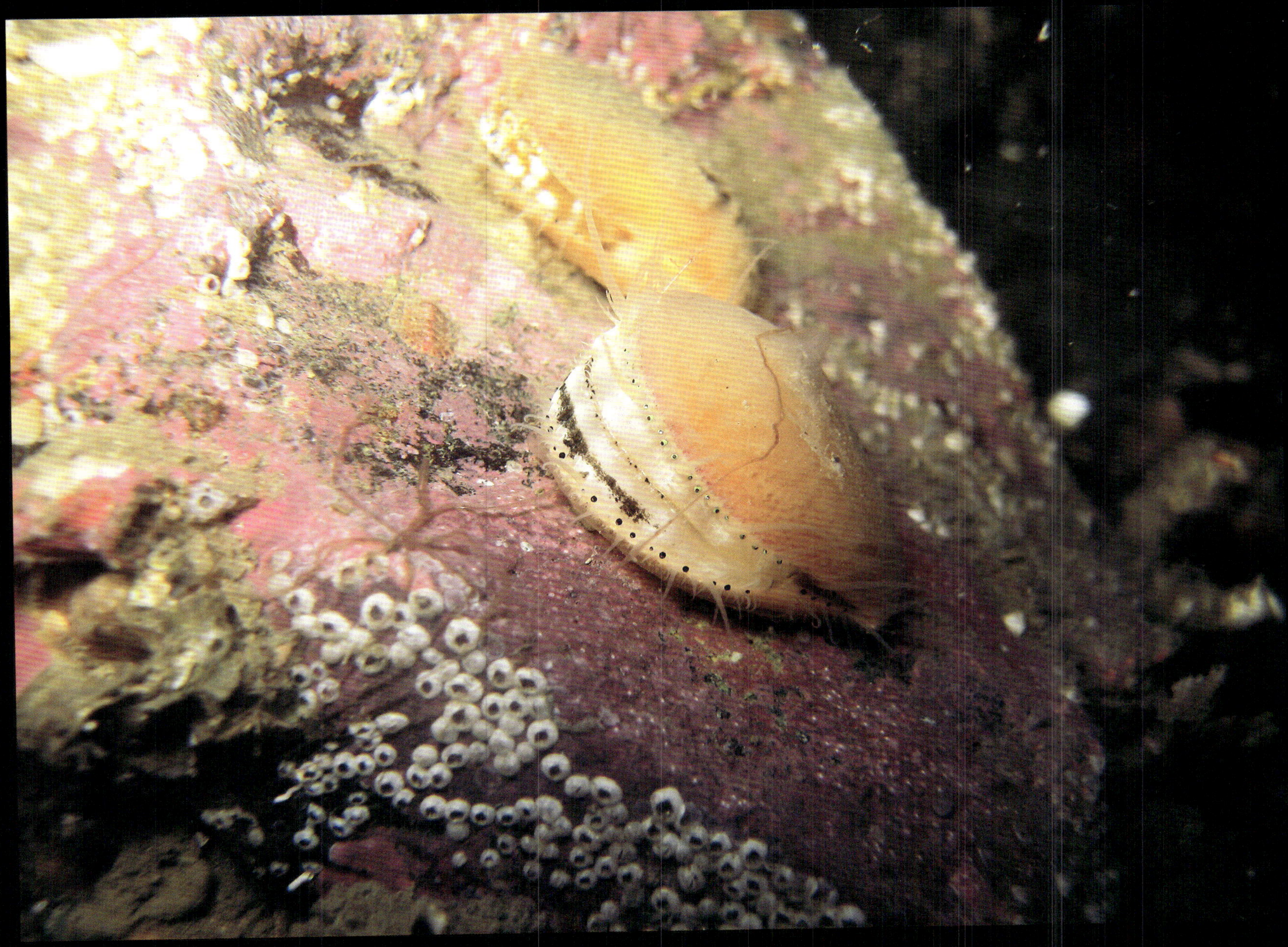

The pink scallop (*Chlamys rubida*) has smooth shell ridges that are often covered by sponges as seen here. When threatened by predators, these bivalves rapidly clap their shells together and "fly" away. The dark dots between the two valves are rudimentary eyes that determine the difference between light and dark. *Bay of Islands, Adak Island, R. Brewer photo.*

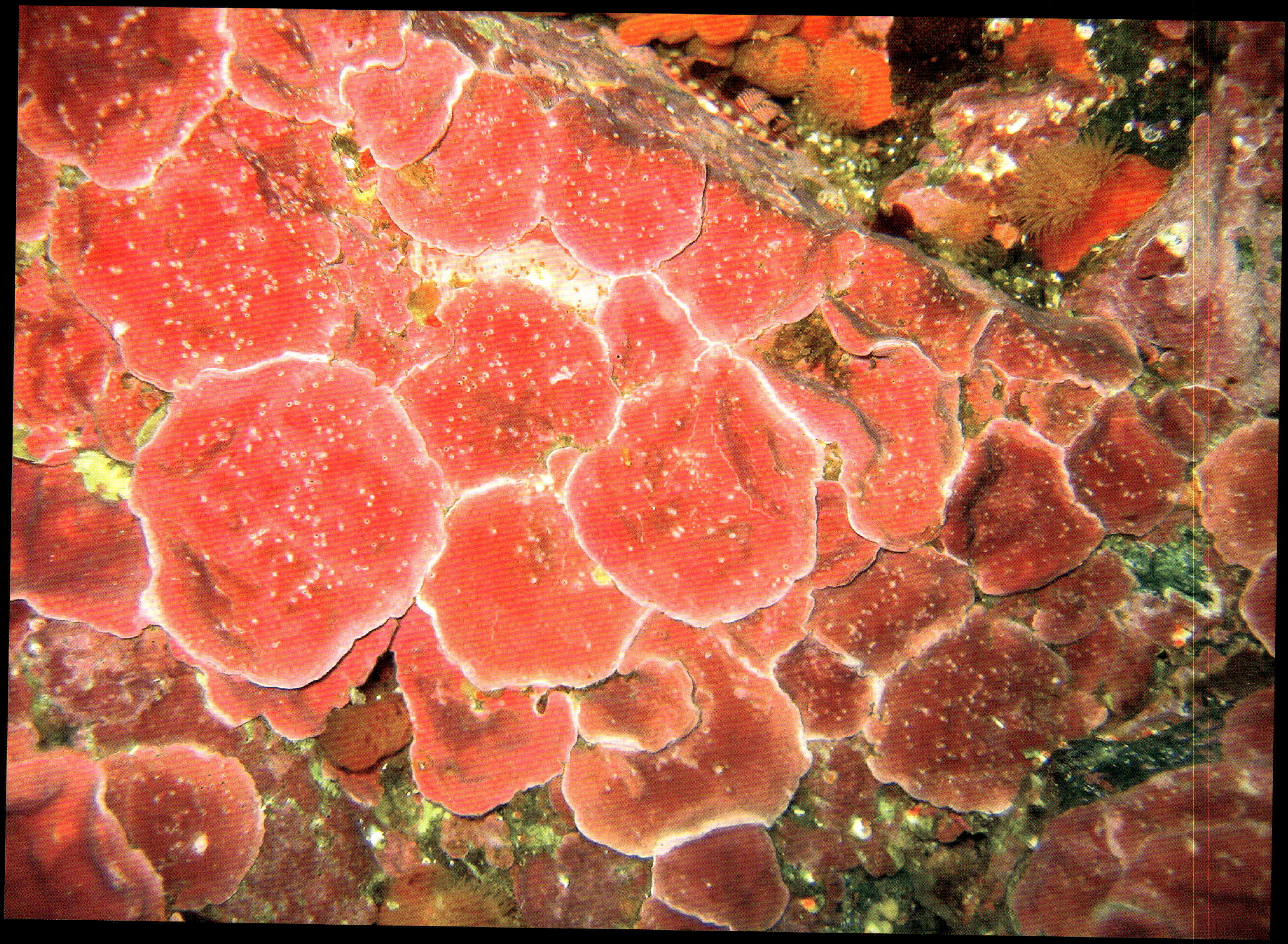

The thick crustose coralline alga (*Clathromorphum nereostratum*) occurs from the Aleutian Islands to the Kuril Islands near Japan. It can reach the age of 700 years, and may be the oldest living alga worldwide. *Clathromorphum nereostratum* forms thick crusts (to 20 in or 50 cm thick) that provide habitat for burrowing organisms. Coralline algae get their stony appearance from the calcium carbonate secreted in their tissue. Red pigments combine with the white carbonate mineral to produce pink coloration. A benefit of calcification is that it deters grazing on coralline algae, but calcification results in extremely slow growth at less than 1 mm per year. Researchers interested in climate change are looking at *C. nereostratum* as an indicator species. *Atka Island, H. Chenelot photo.*

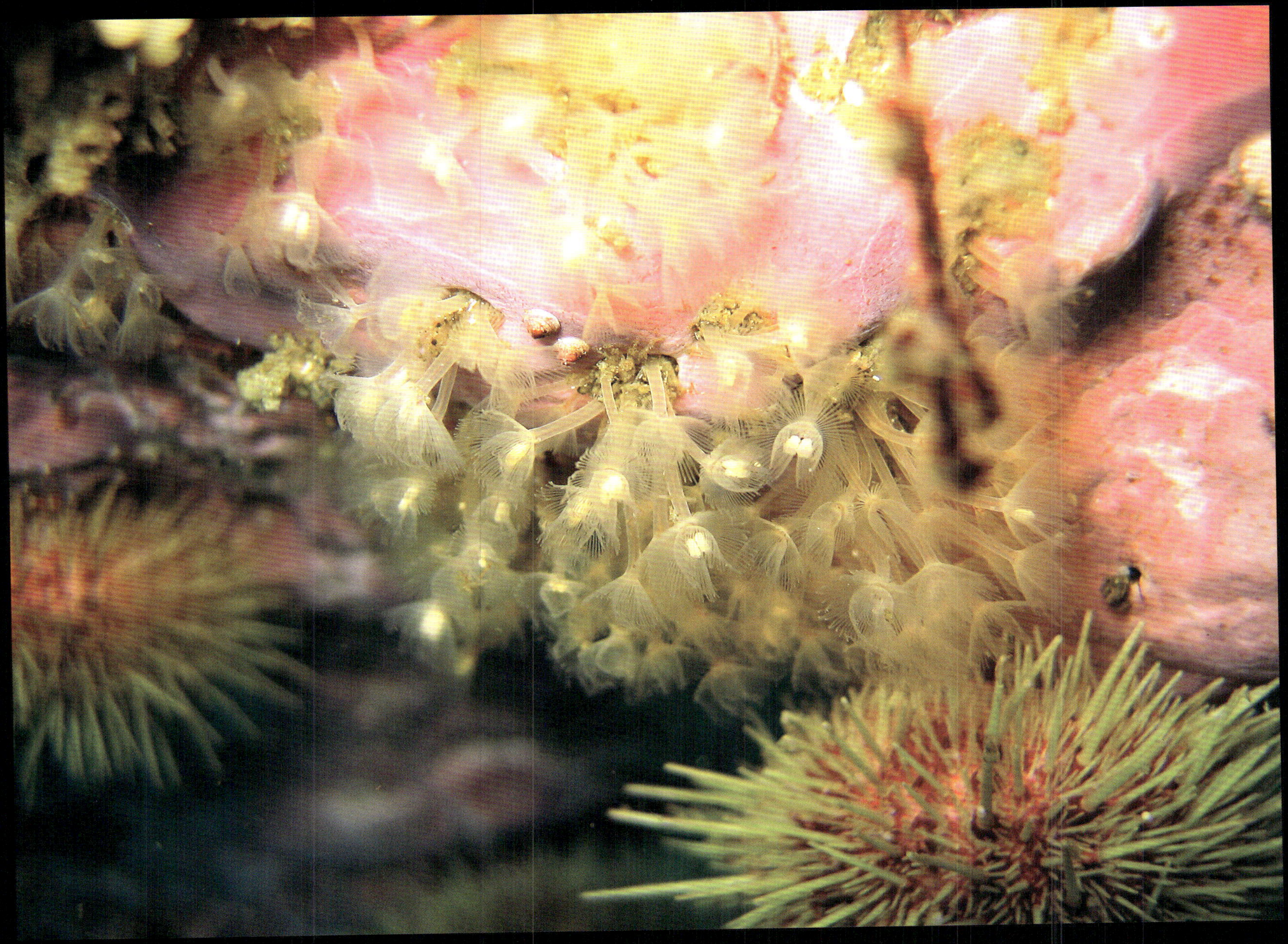

Phoronids have a wormlike body and a horseshoe-shaped corona (crown). They are filter-feeders with a U-shaped digestive tract that begins and ends near the corona and mouth. Six species of phoronids inhabit marine waters of the Pacific Northwest. *Bay of Islands, Adak Island, H. Chenelot photo.*

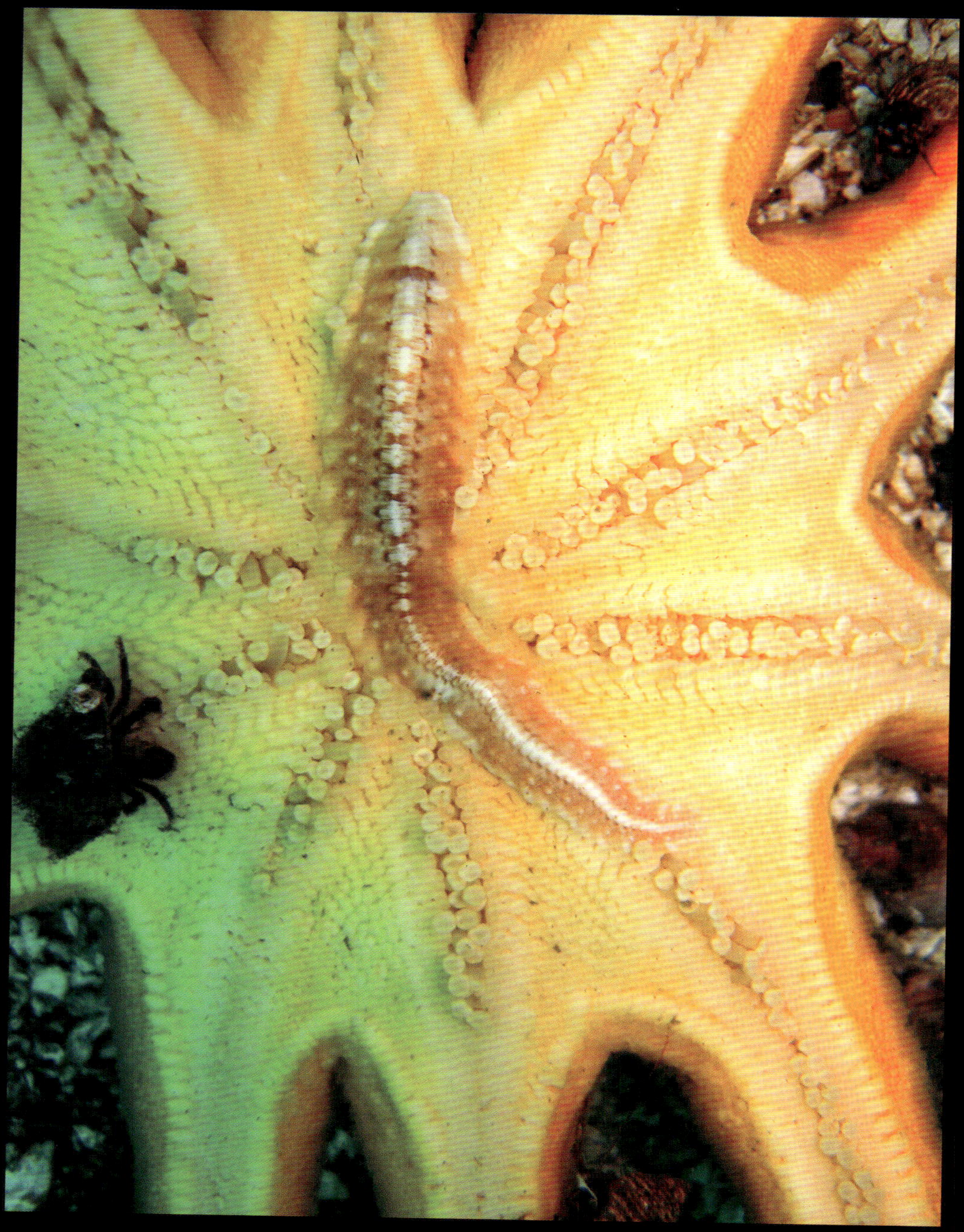

The red-banded commensal scaleworm (*Arctonoe vittata*) is mutualistic with various species of sea stars, snails, and sea cucumbers. As the worm cleans and defends its host, the host provides food and a mobile place to live. *Arctonoe vittata* is fiercely territorial and will even attack other members of its own species. *Kagalaska Island, S. Jewett photo.*

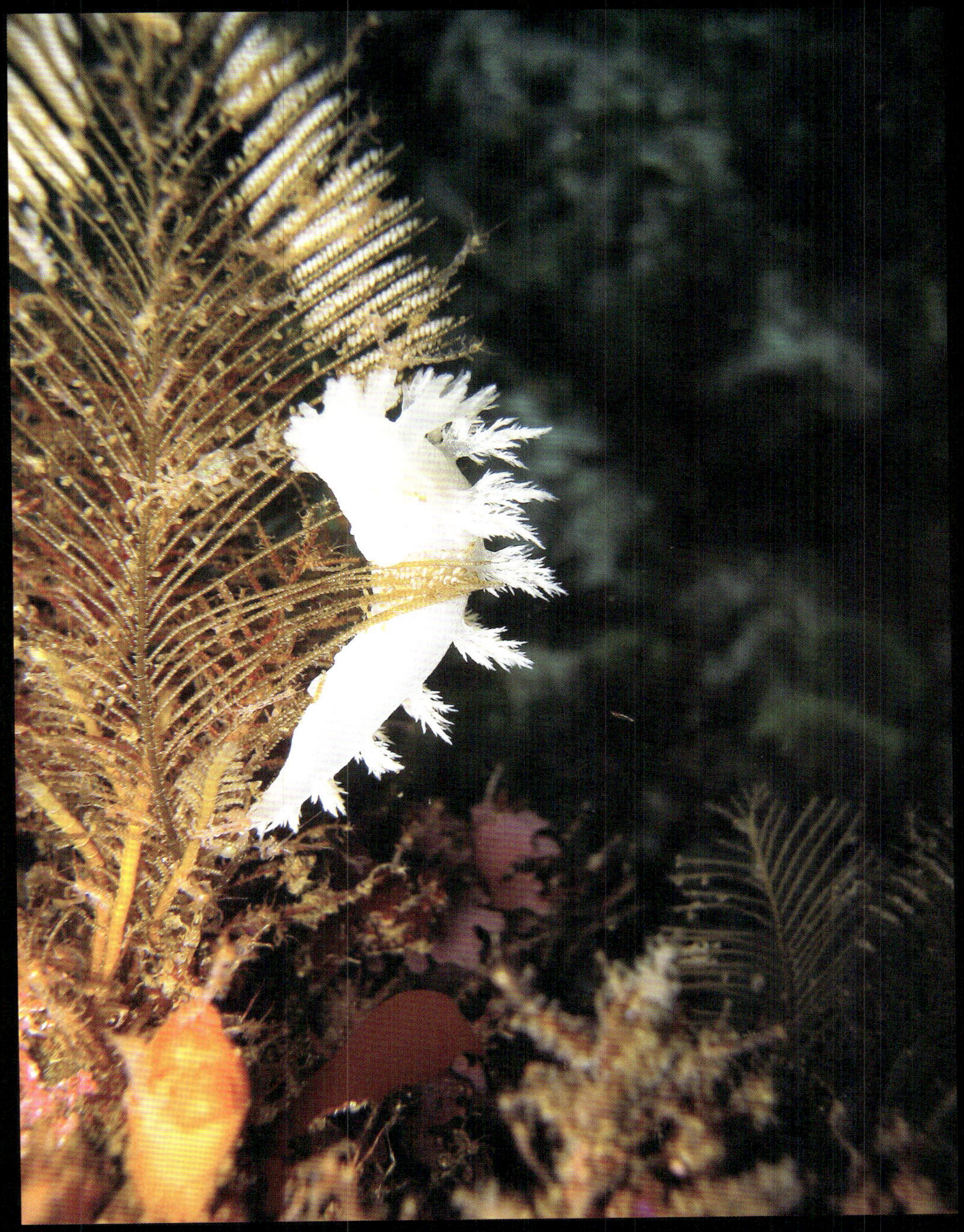

Dall's dendronotid (*Dendronotus dalli*) has paired, white-tipped gills along its back and is a capable swimmer when dislodged from the surface it lives on. It eats hydroids, tube-dwelling sea anemones, and stalked bryozoans. Dall's dendronotid was named to honor William Healey Dall, a prominent mollusk researcher, and one of the earliest scientific explorers of interior Alaska. *Deep Bay, Atka Island, R. Brewer photo.*

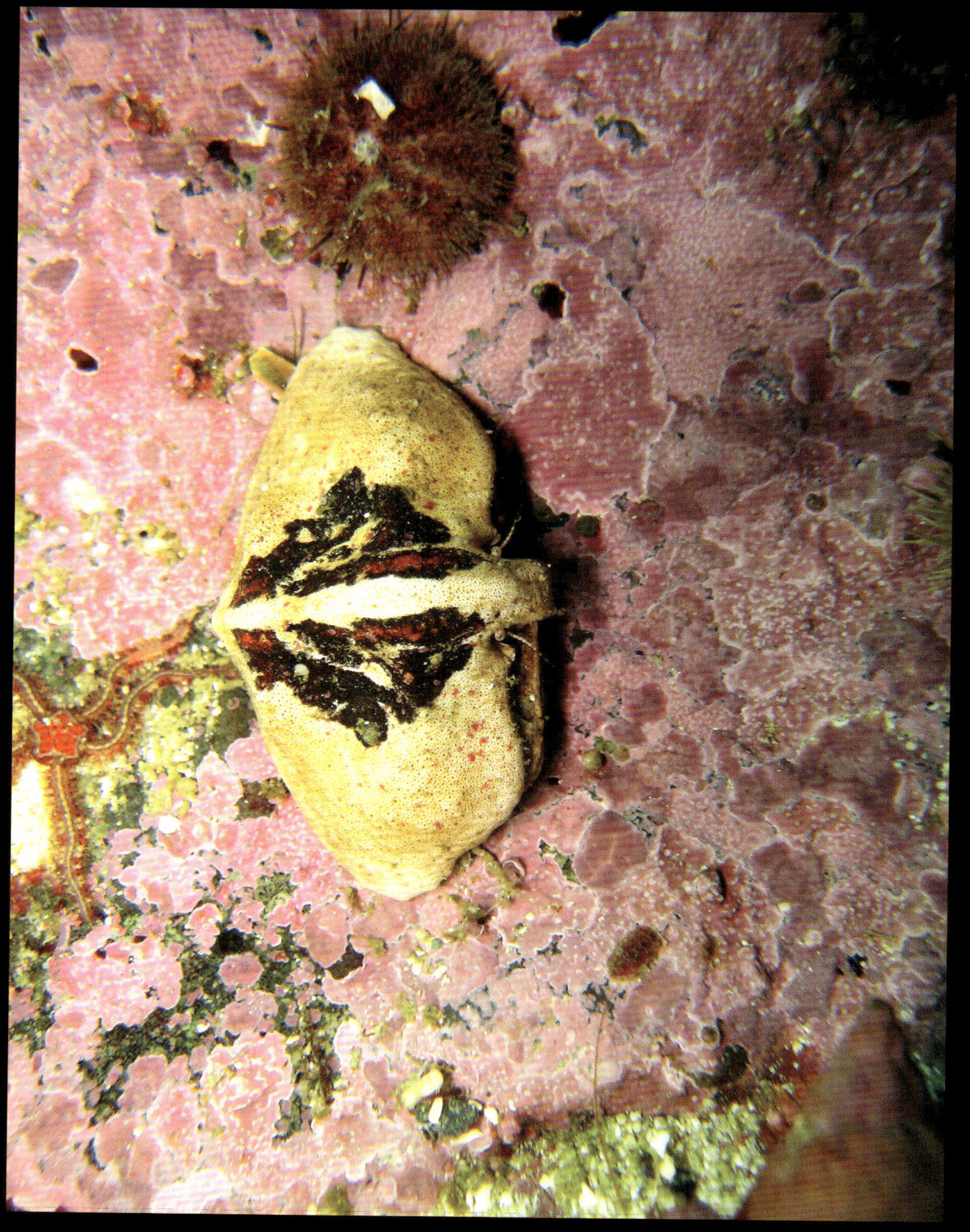

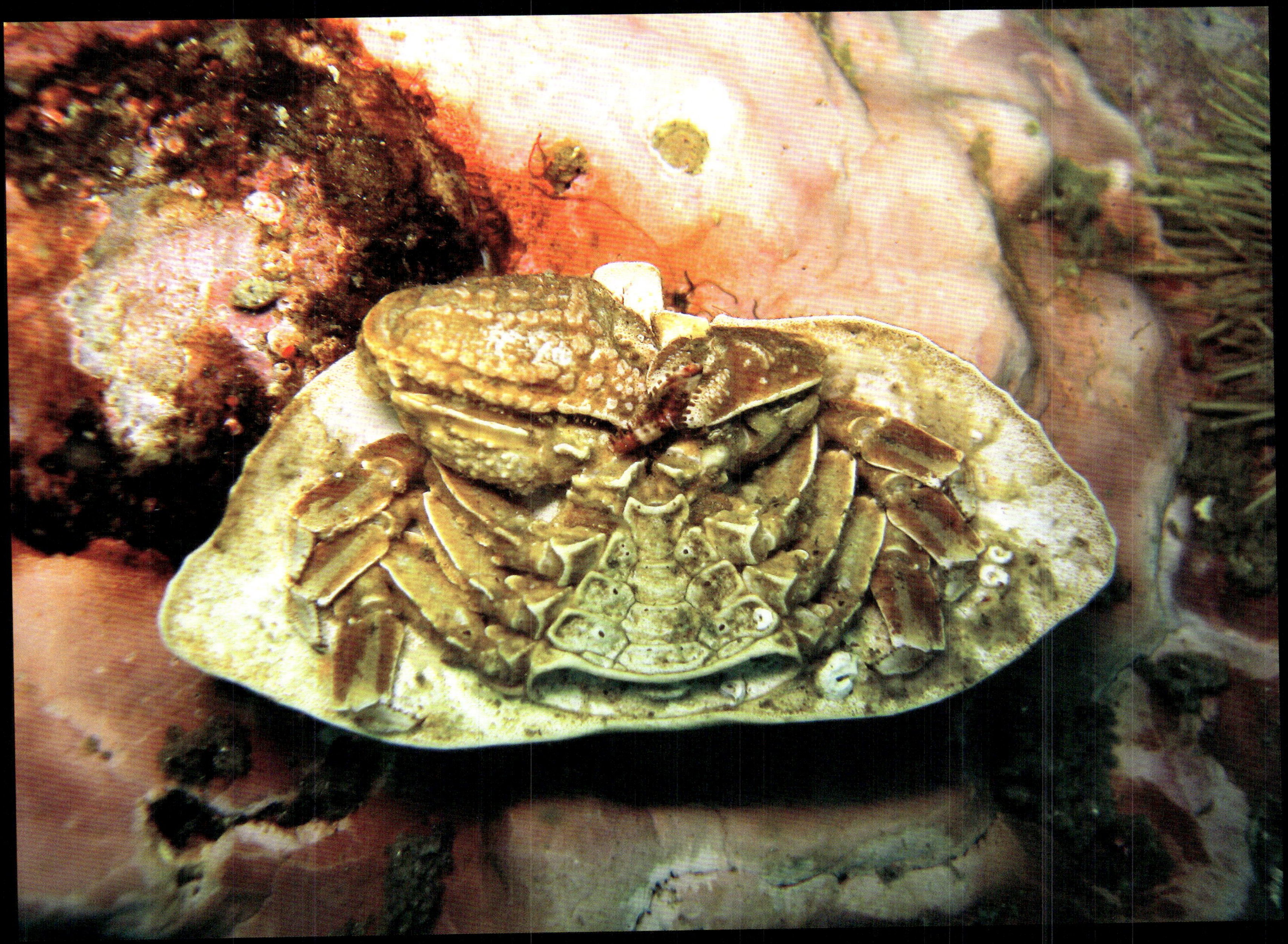

The butterfly crab (*Cryptolithodes typicus*) comes in a wide variety of colors and is often difficult to detect unless it moves. The butterfly crab grazes on bryozoans, coralline algae, and other organisms attached to seafloor surfaces. Often found among rubble at the base of steep, rocky slopes, this small crab occurs from Santa Rosa, California, to Amchitka Island, Alaska. *Kuluk Bay, Adak Island, S. Jewett photos.*

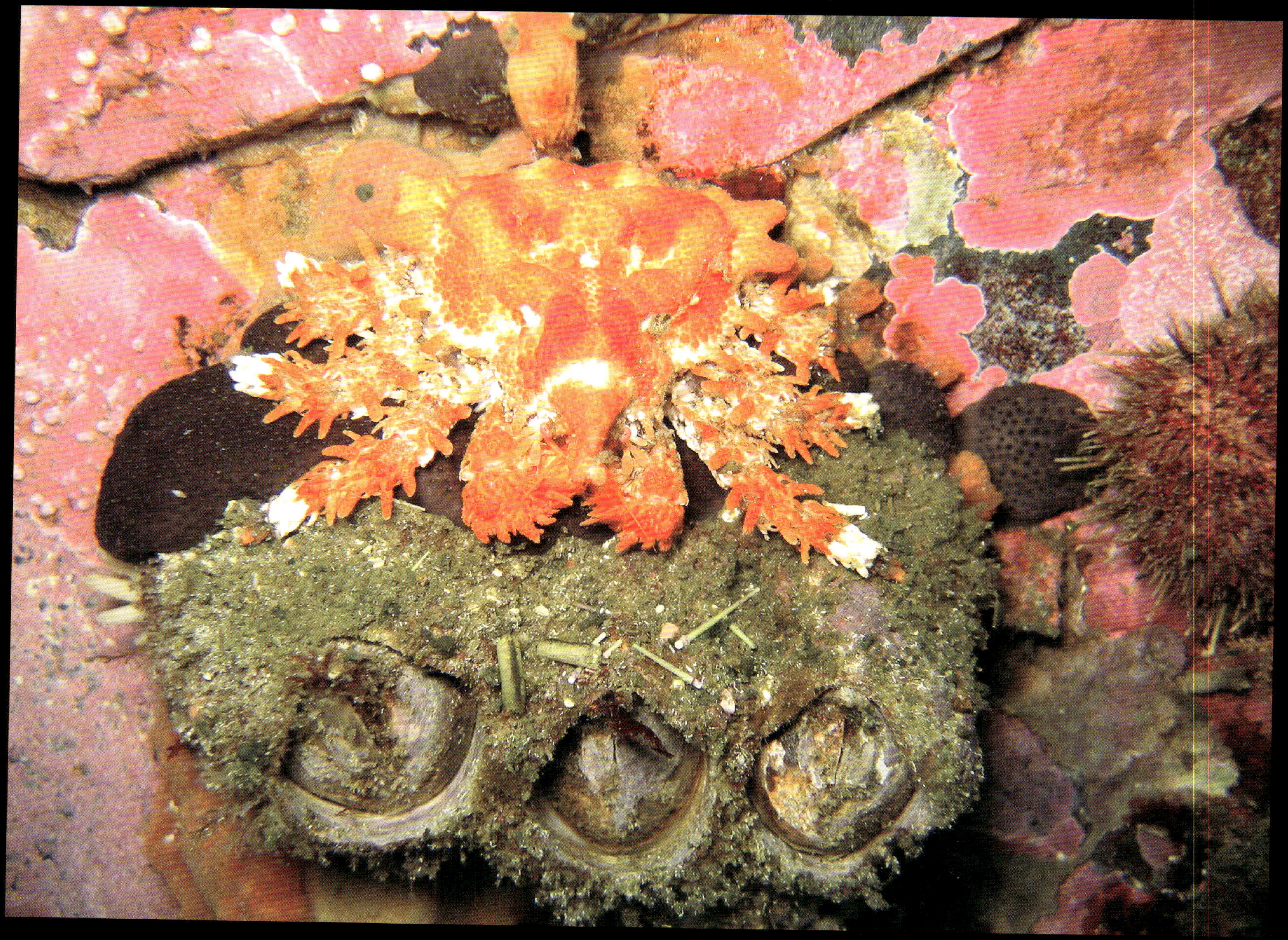

A heart crab (*Phyllolithodes papillosus*) is identified by its triangular carapace and raised, heart-shaped back. Although juveniles of this species inhabit the intertidal, most adults live in subtidal areas down to 590 ft (180 m). Individuals are well camouflaged and usually are seen only when they move. *Finger Bay, Adak Island, H. Chenelot photo.*

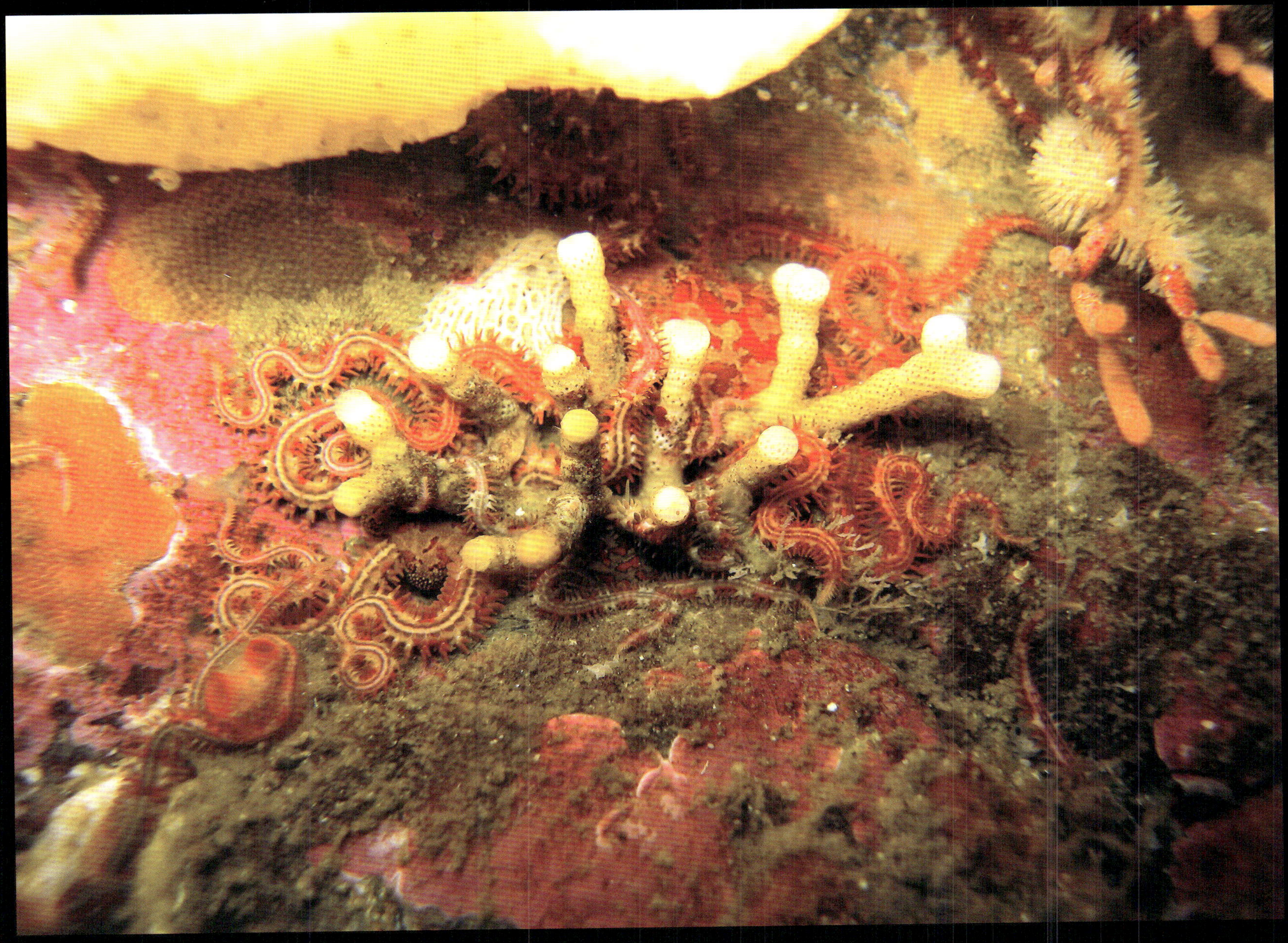

The northern staghorn bryozoan (*Heteropora alaskensis*) has a calcified skeleton and can form dense clumps of coral-like projections. Bryozoans are colonies of small, near-independent zooids that filter-feed using millions of tiny cilia or hairs. This colony is hiding some daisy brittle stars. Bryozoans are called "moss animals" because the encrusting bryozoans look like patches of moss. *Bay of Islands, Adak Island, R. Brewer photo.*

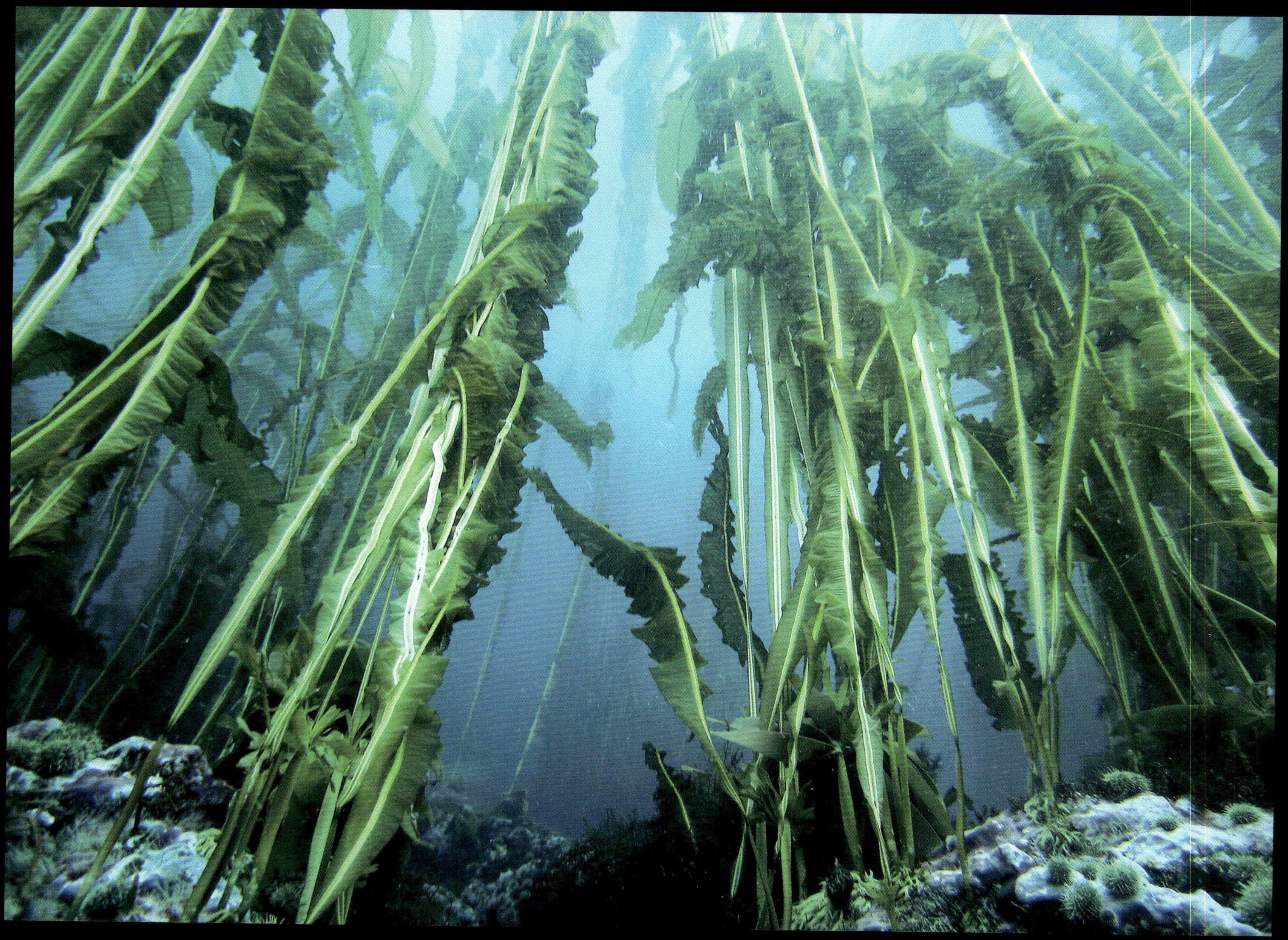

Dragon kelp (*Eualaria fistulosa*, formerly *Alaria fistulosa*) is a brown alga that forms a canopy using gas-filled bladders to reach 80 ft (25 m). A single blade can be 6 ft (2 m) wide. This species provides a complex habitat for fishes and invertebrates and is usually found in rocky areas with some surf or current. *Cape Idalug, Amlia Island, R. Brewer photo.*

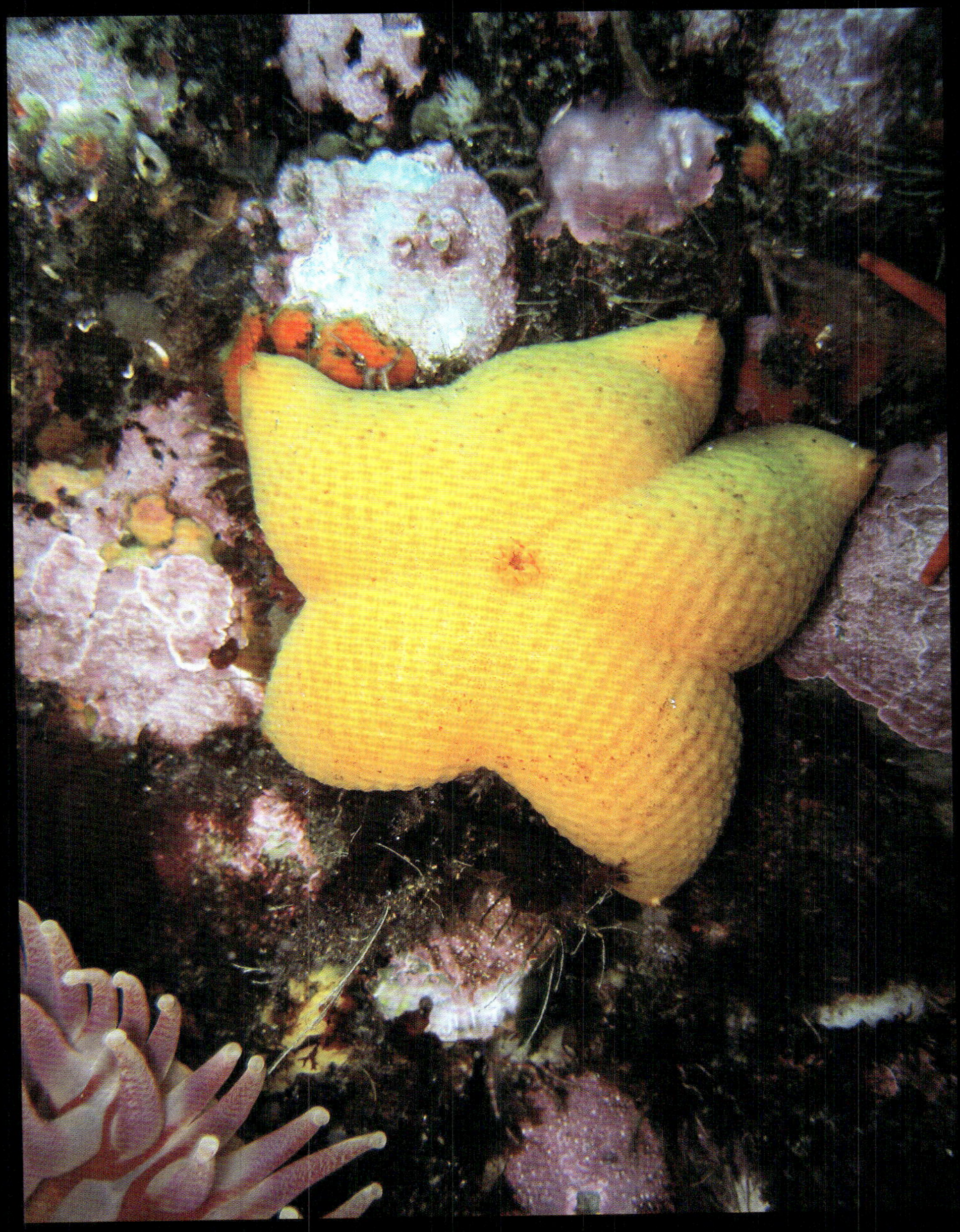

The cushion star or slime star (*Pteraster tesselatus*) has distinctively short arms and a puffy appearance, but is best known for its ability to produce large amounts of slime as a defensive response. This species feeds on sponges and is found from California to Alaska to Japan. *Cape Idalug, Amlia Island, R. Brewer photo.*

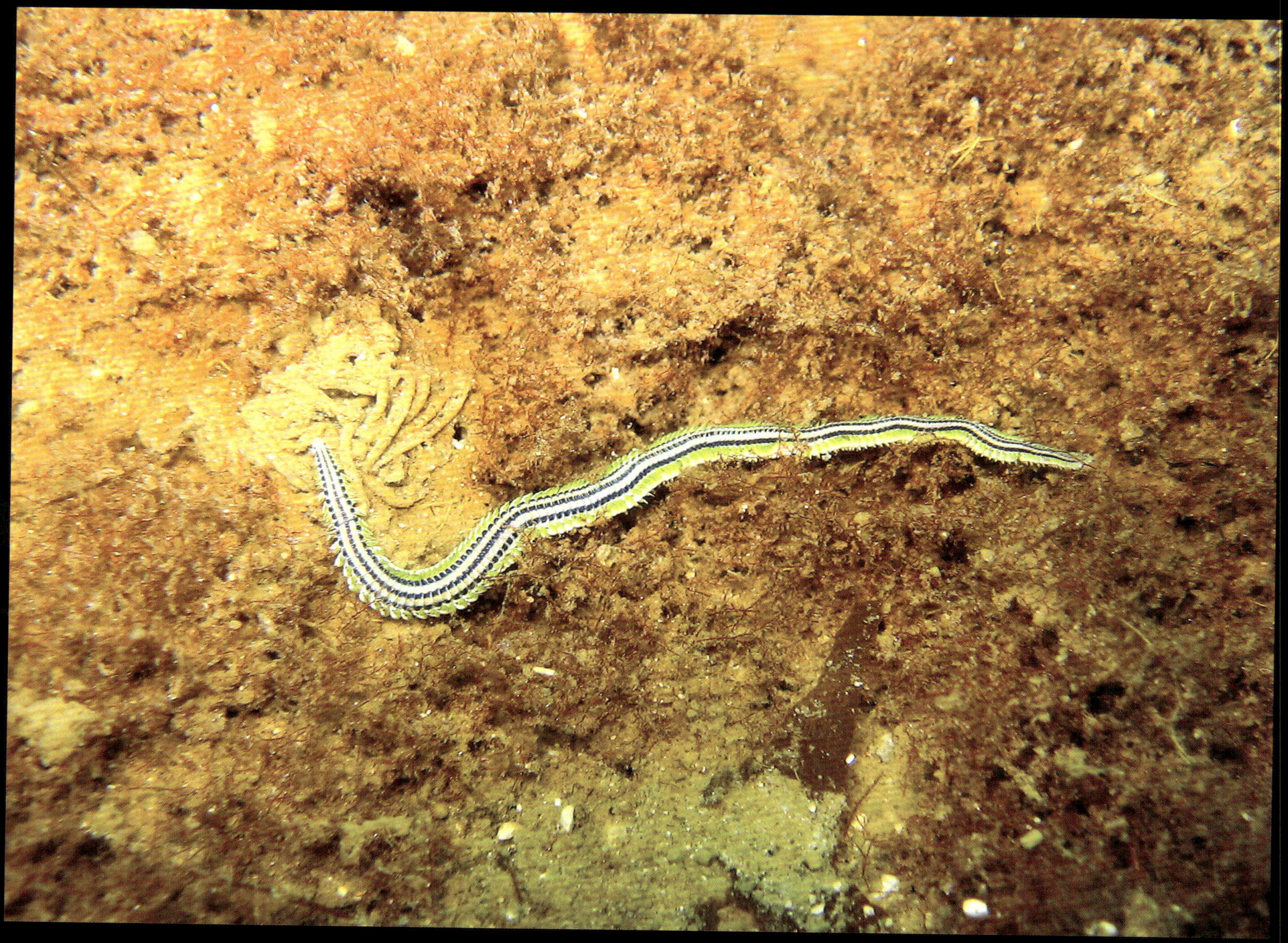

Paddleworms (*Eteone* sp., Phyllodocidae) are common shallow-water polychaetes, usually associated with hard substrates. Paddleworms flutter their dorsal cirri (projections) to aid in oxygen exchange. Most are carnivores and some hunt by following mucous trails left by gastropods. *Chisak Bay, Little Tanaga Island, H. Chenelot photo.*

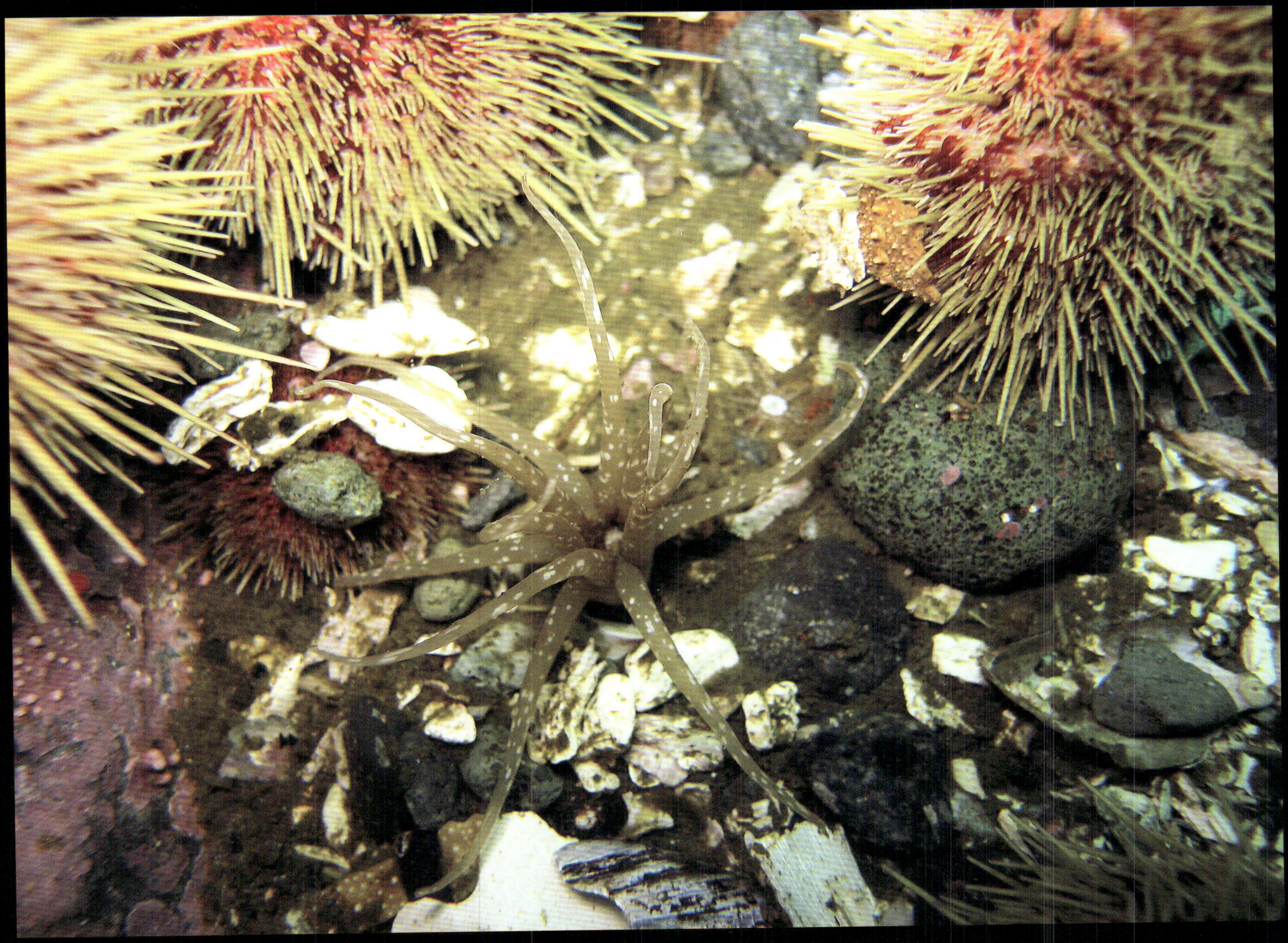

This common 12-tentacle anemone burrows into coarse sand/shell hash or between large rocks. Sea anemone expert Meg Daly puts this anemone in the family Edwardsiidae. Like most anemones, it probably is a planktonic feeder that waits for prey to touch the nematocysts (stinging cells) that line its arms, and then sweeps food past a one-way gut. *Bay of Islands, Adak Island, R. Brewer photo.*

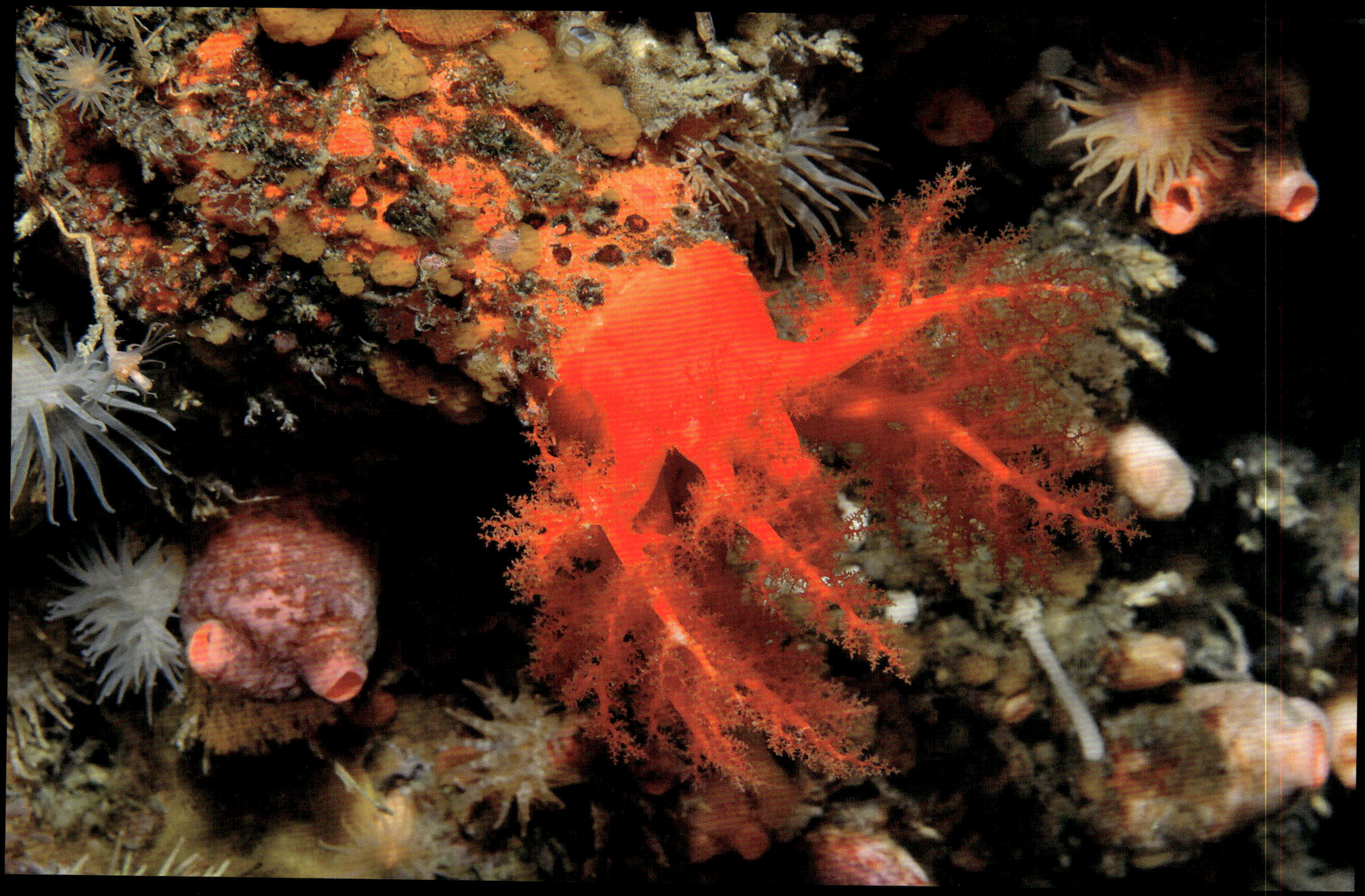

The armored sea cucumber (*Psolus fabricii*) has stiff shingle-like scales on the dorsal surface with a burnt orange color. This species is found in areas with large boulders and high water flows. Similar to other echinoderms, it has a water-based skeleton with rows of tubefeet. *Nazan Bay, Atka Island, S. Harper photo.*

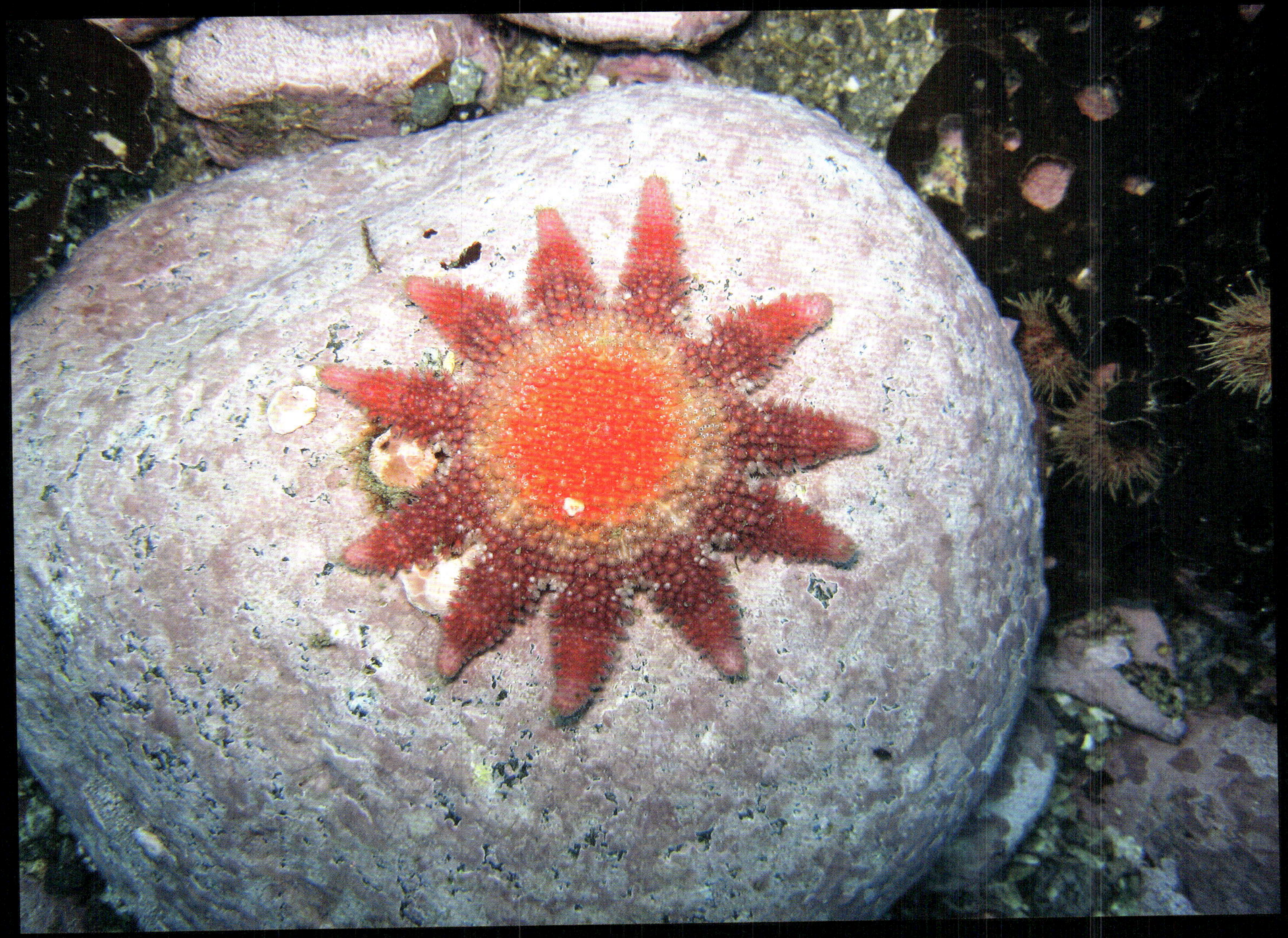

The rose star (*Crossaster papposus*) has a distinctive radiating color pattern of red, orange, and white. This species usually has 11 rays (arms) and feeds on bivalves, ascidians, nudibranchs, and other sea stars. The rose star has a circumpolar range and lives on soft and hard surfaces from the low intertidal to 4,000 ft (1,200 m) depth. *Umak Pass, Little Tanaga Island, R. Brewer photo.*

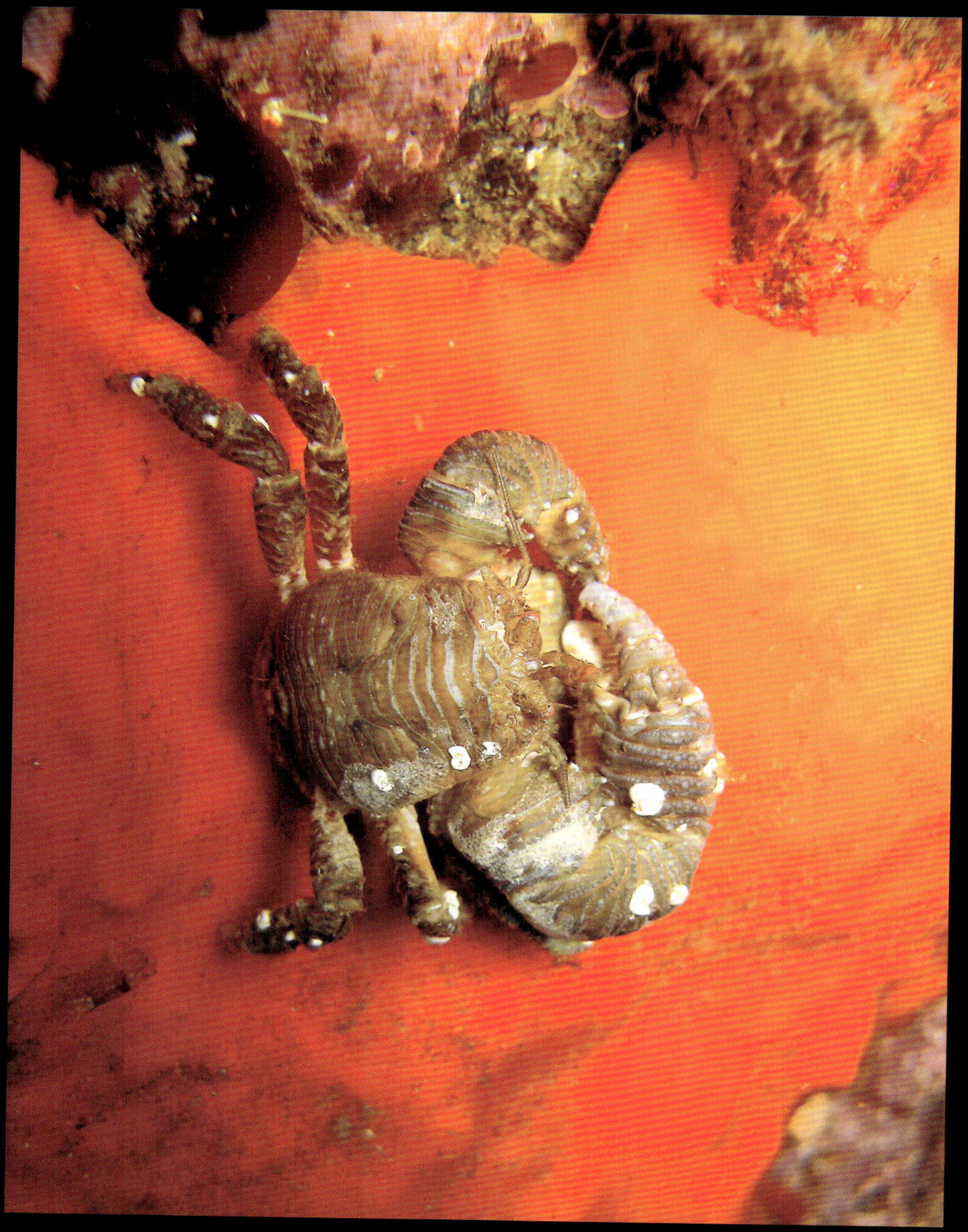

The wrinkled crab (*Dermaturus mandtii*) is easily recognizable by its ring pattern and unequal claw size. This species is found among kelp holdfasts and under rocks, and feeds on algae or algal detritus. The wrinkled crab frequents the low intertidal to 240 ft (73 m) from Alaska to Japan. *Deep Bay, Atka Island, R. Brewer photo.*

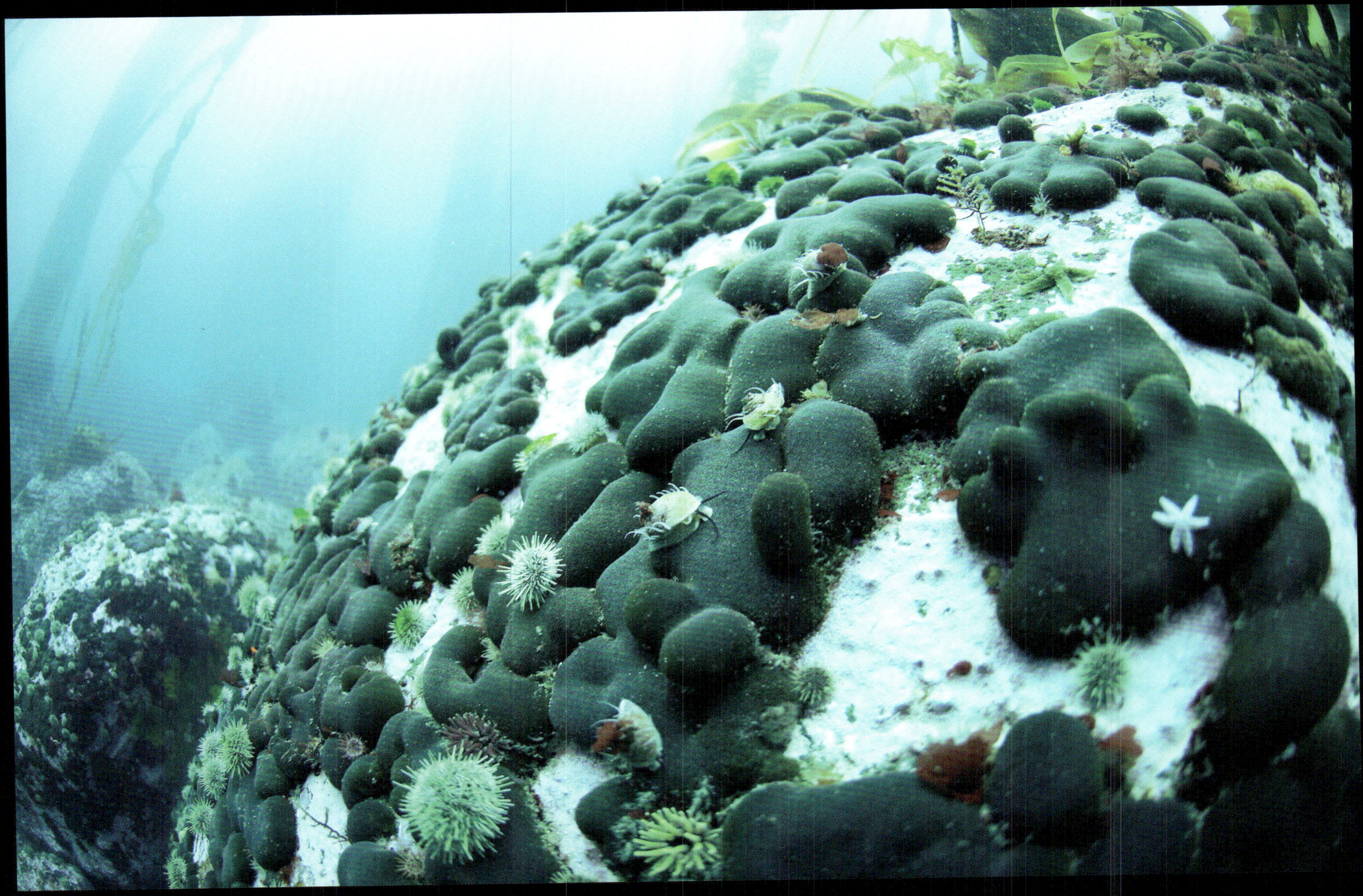

The green alga Ritter's spongy cushion (*Codium ritteri*) is found throughout the Aleutian Island chain, in some areas covering much of the landscape. Each cushion (to 4 in or 10 cm diameter) is a single multinucleate cell. Species of *Codium* are important food for certain nudibranchs; intact chloroplasts from the algae continue to photosynthesize after being eaten. *Vasilief Bay, Atka Island, S. Harper photo.*

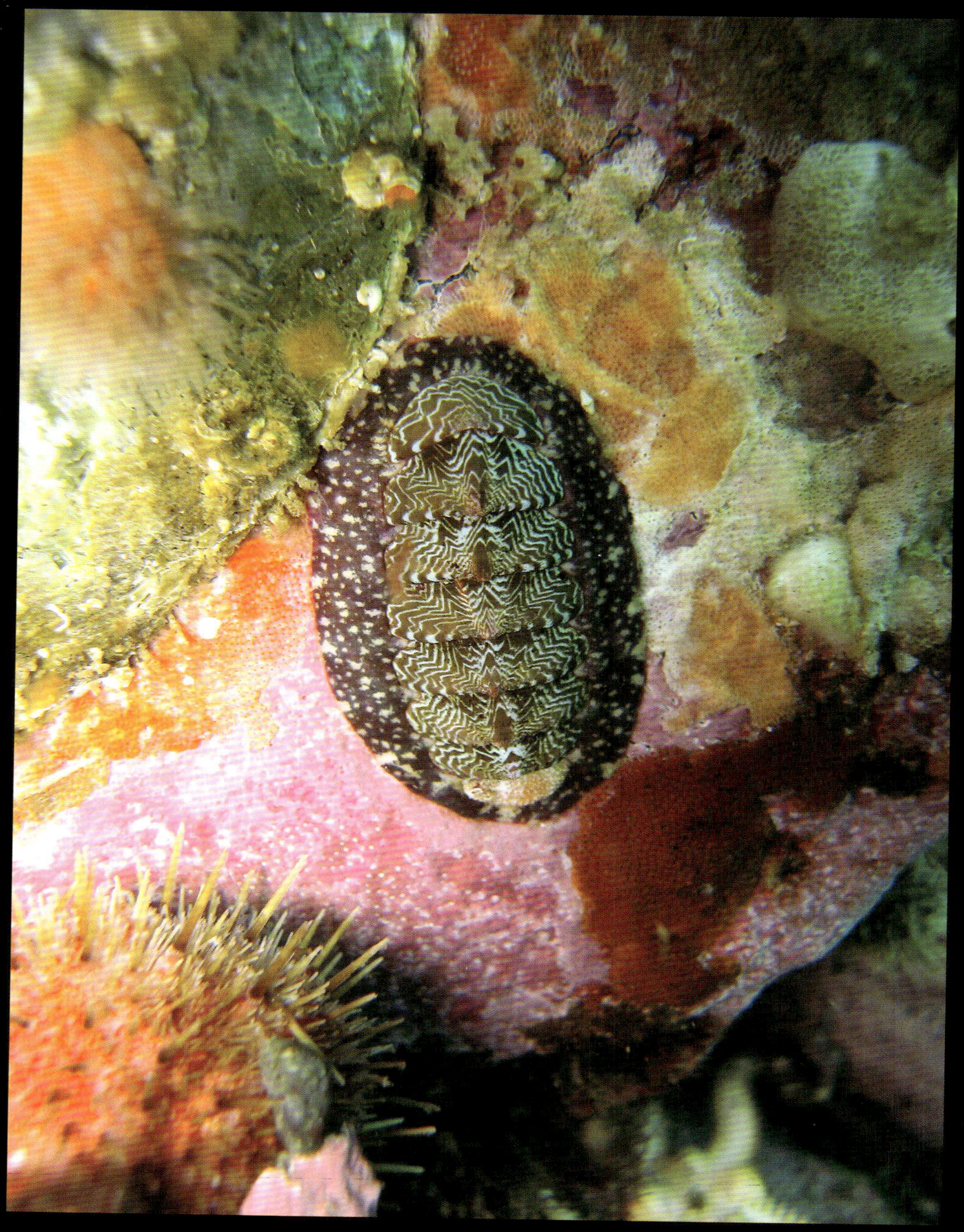

The lightning chiton (*Tonicella submarmorea*) has distinctive white spots on its mantle and white zigzag patterns on its valves (shells). Chitons are in a group of mollusks called the polyplacophorans, which means "many shells." Instead of having a complete shell like a snail, chitons bear eight shells that protect and provide structure. This species is found throughout the central and western Aleutians. *Umak Bight, Umak Island, R. Brewer photo.*

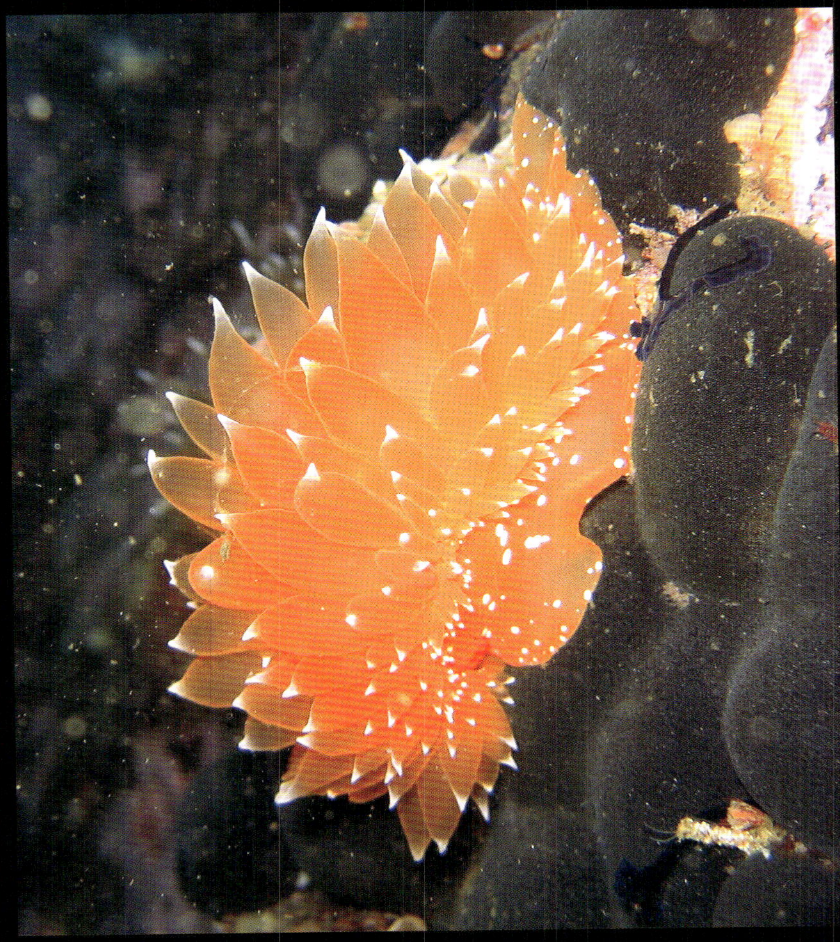

The golden dirona (*Dirona pellucida*) is distinguished from other dironas by the white speckling in its lower body. The large paddle-like structures on the back are external gills where oxygen and carbon dioxide exchange occurs. This species is found on rocky bottoms at depths less than 150 ft (50 m). *Finger Bay, Adak Island, H. Chenelot photo.*

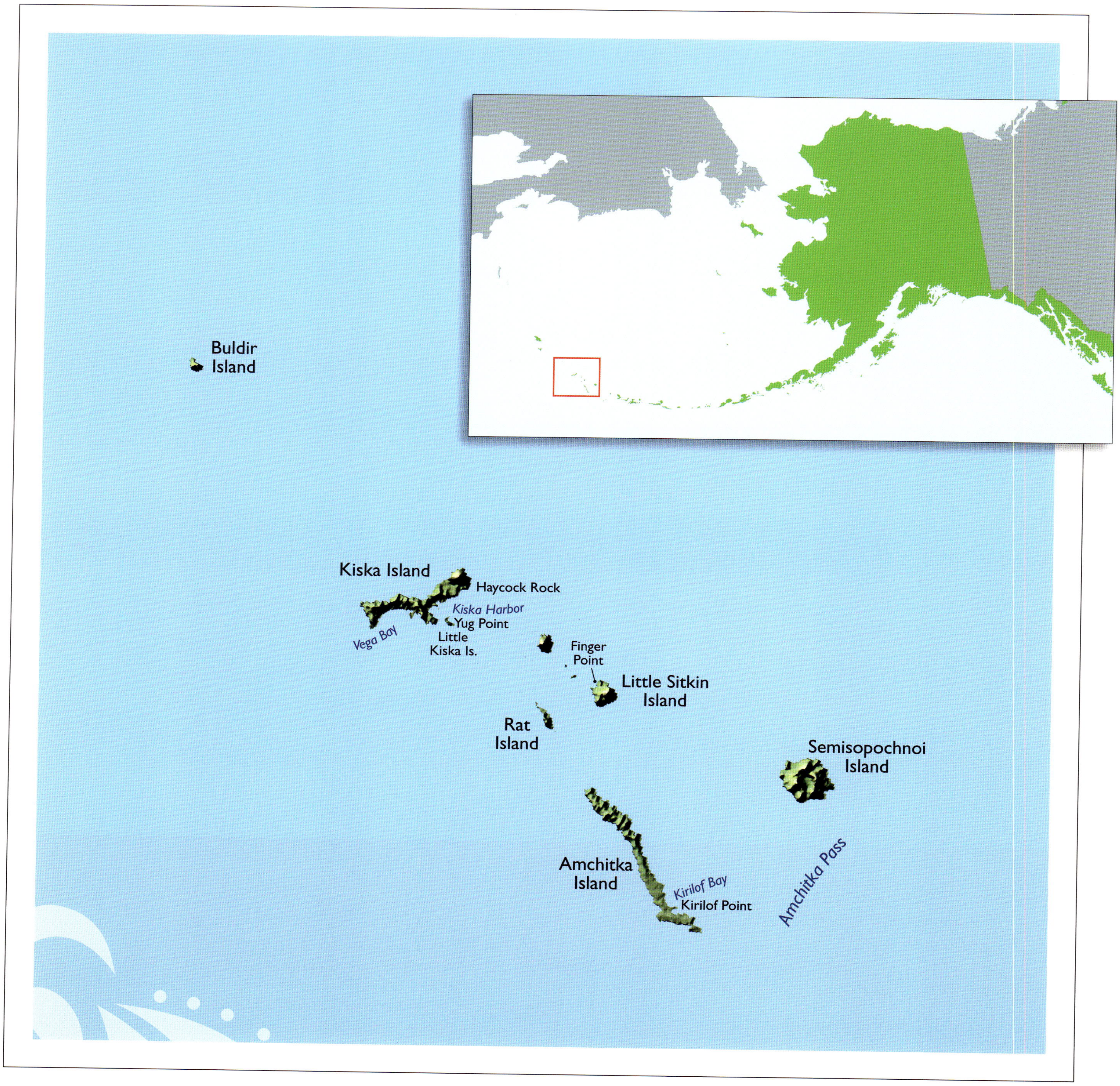
Buldir
Island
Kiska Island
Haycock Rock
Kiska Harbor
Yug Point
Little
Kiska Is.
Vega Bay
Finger
Point
Little Sitkin
Island
Rat
Island
Semisopochnoi
Island
Amchitka
Island
Kirilof Bay
Kirilof Point
Amchitka Pass

Rat Islands

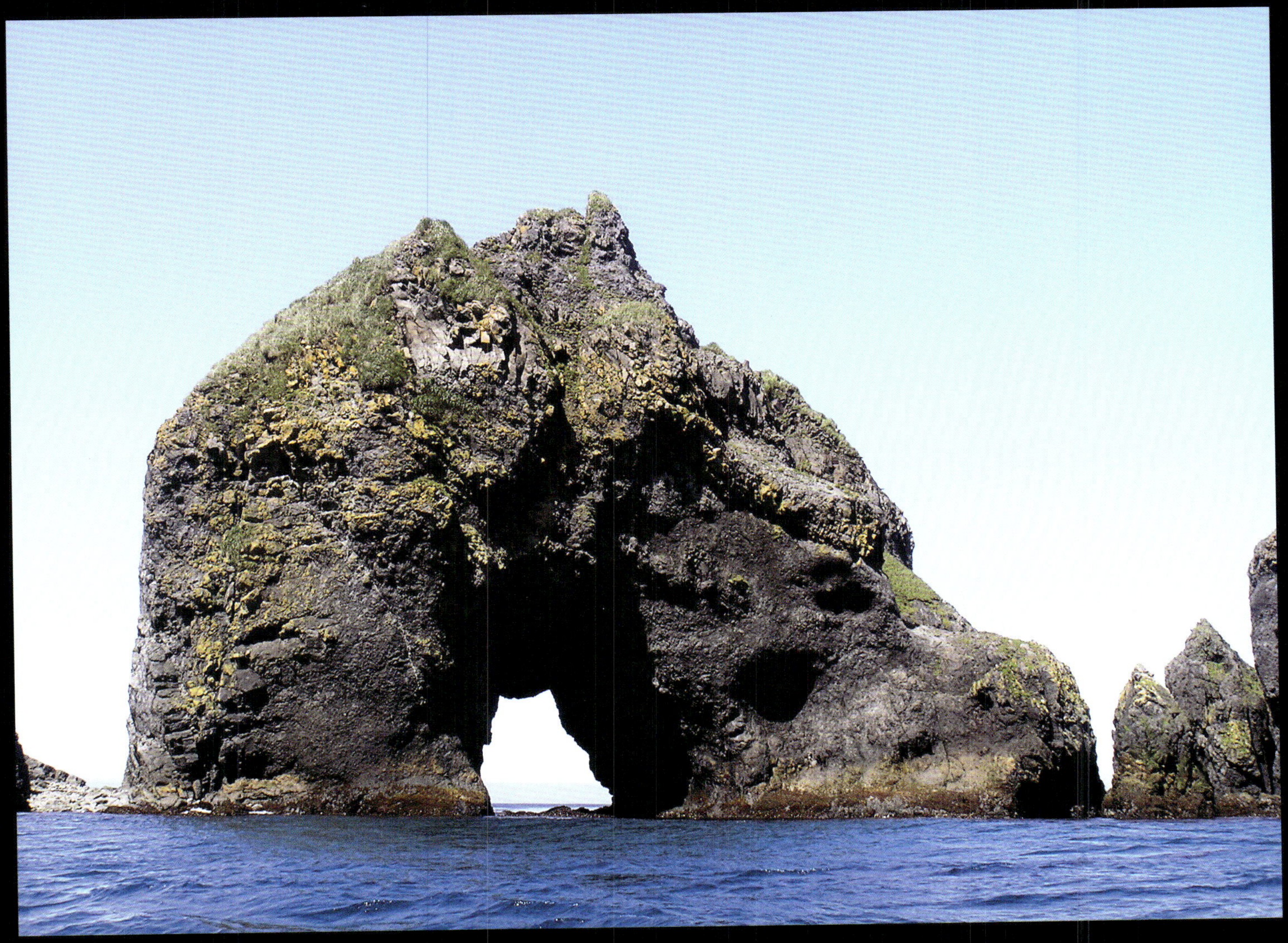

Geological formations like this isolated arch are fairly common throughout the Aleutian Islands. In exposed areas, many of these rock features are eroded by the relentless forces of the ocean. *Kiska Island, J. Burger photo.*

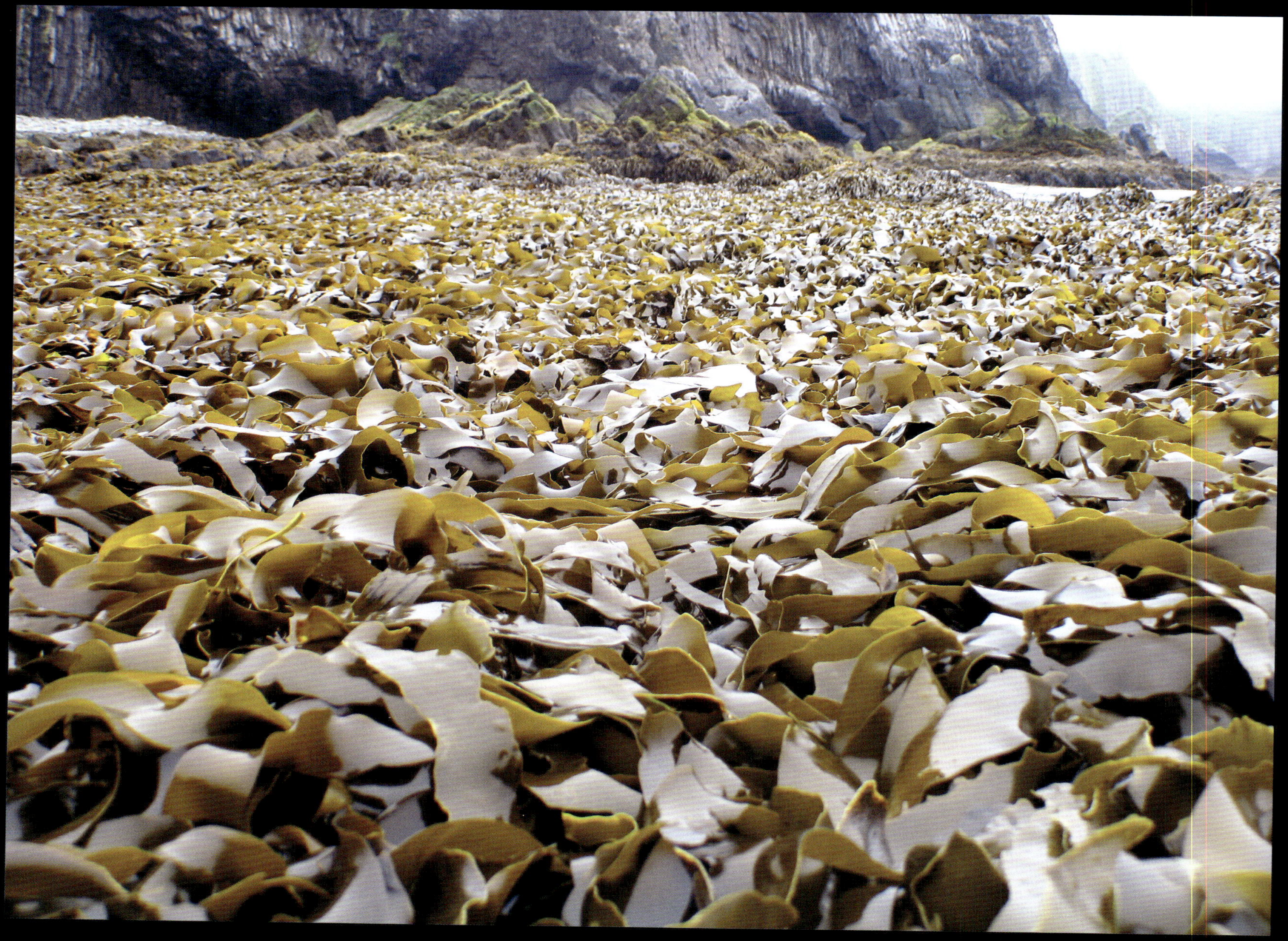

An extremely low tide on Kiska Island reveals a flat of intertidal laminarian kelp. When exposed to air intertidal kelp can lose most of its water through evaporation, and recover completely when the tide returns. Light reflections on the kelp fronds almost appear white. *Vega Bay, Kiska Island, M. Lindeberg photo.*

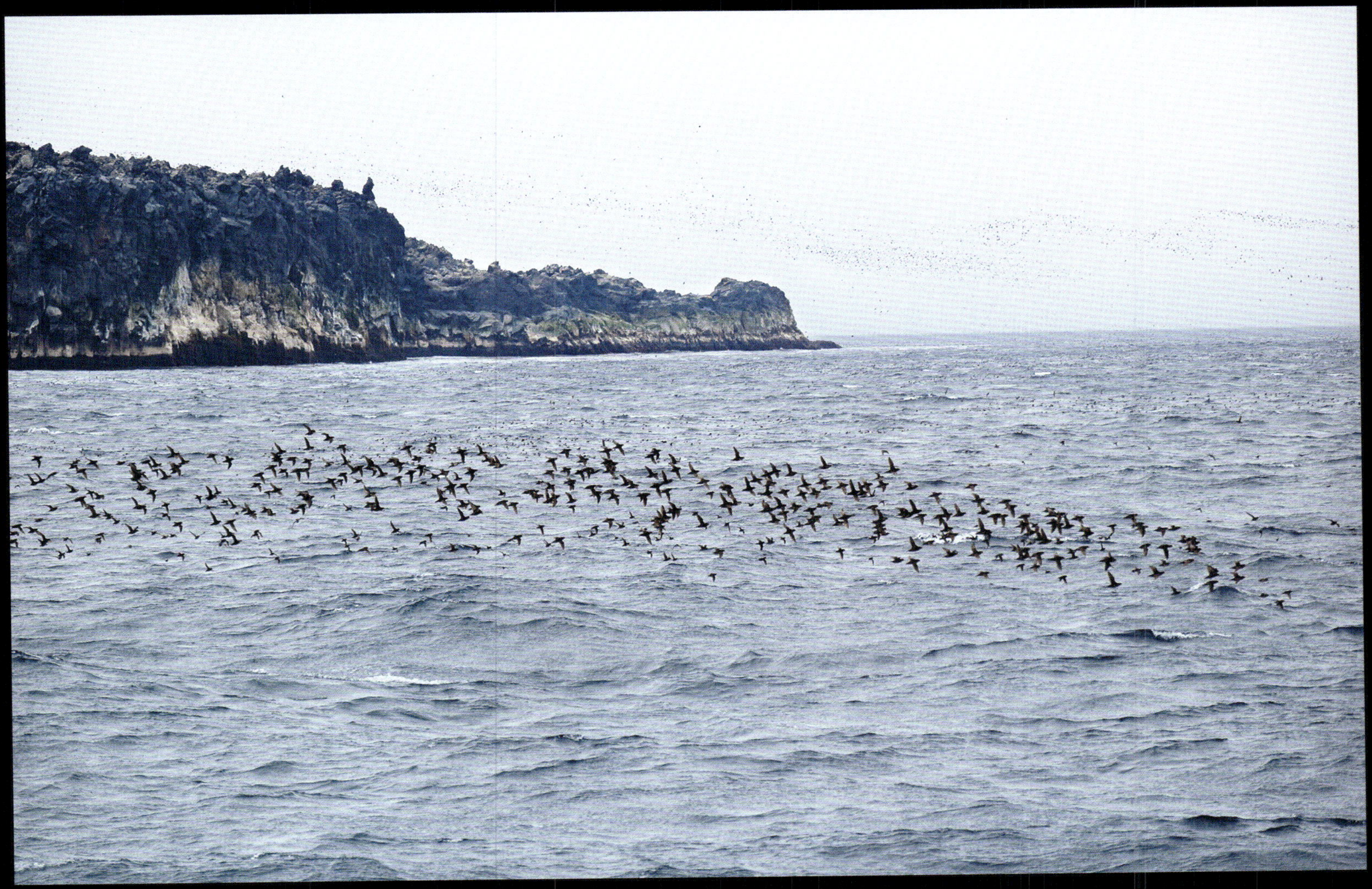

More than a million least auklets (*Aethia pusilla*) and crested auklets (*A. cristatella*) nest on Kiska Island each summer and undertake daily foraging bouts near the island. The solidified lava from the nearby Kiska Volcano provides ideal nesting habitat for the auklets. *Kiska Island, M. Lindeberg photo.*

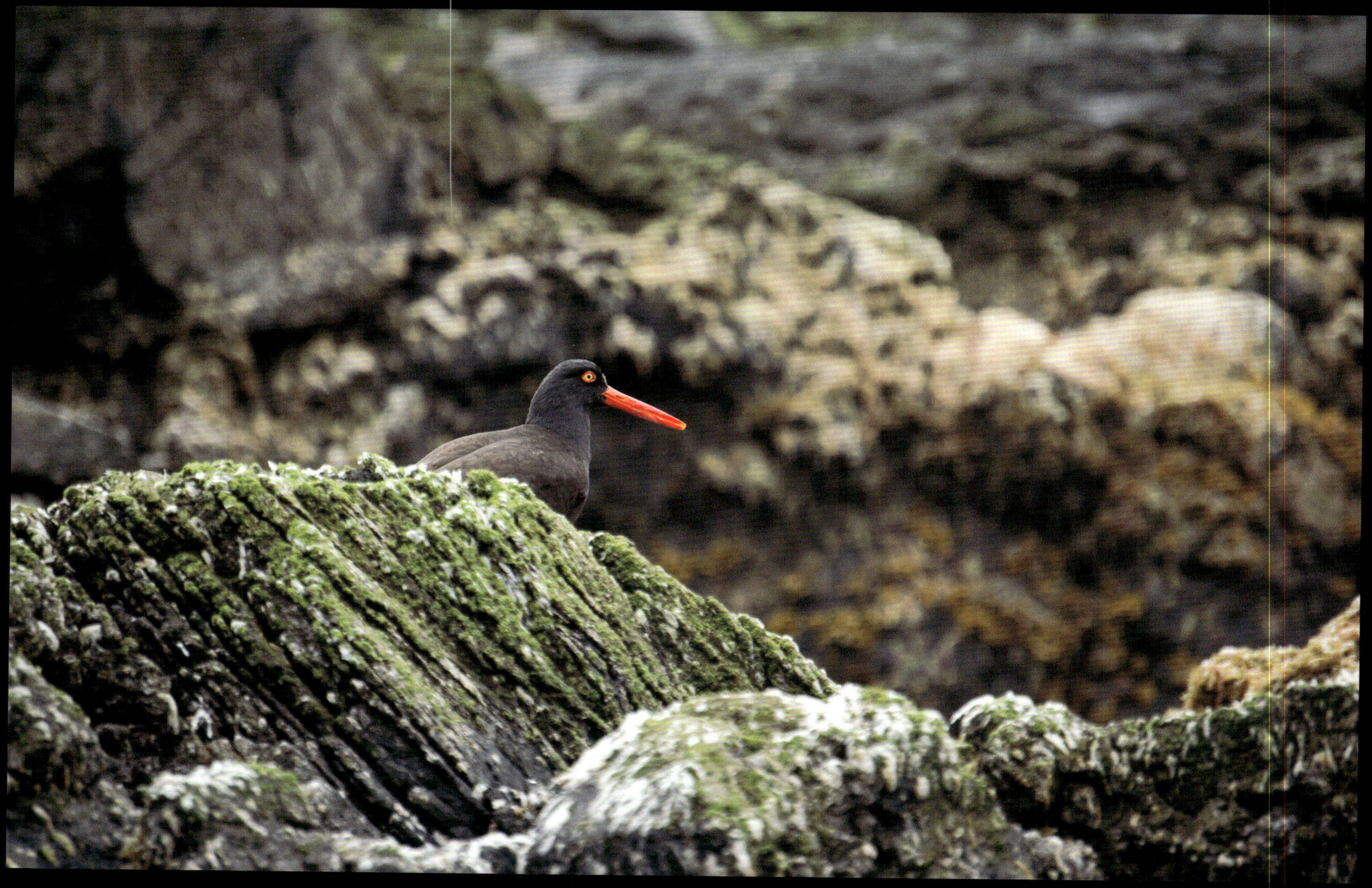

The black oystercatcher (*Haematopus bachmani*) is easily identified by its bright red, elongated bill and its stocky black body. Black oystercatchers are nonmigratory birds that feed on mussels, whelks, and limpets in the intertidal region. *Amchitka Island, M. Lindeberg photo.*

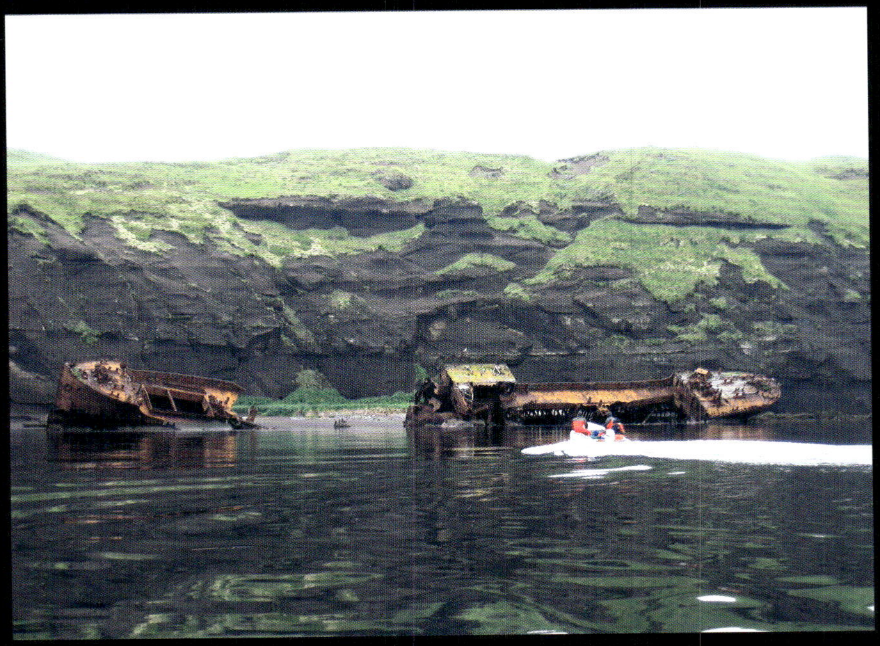

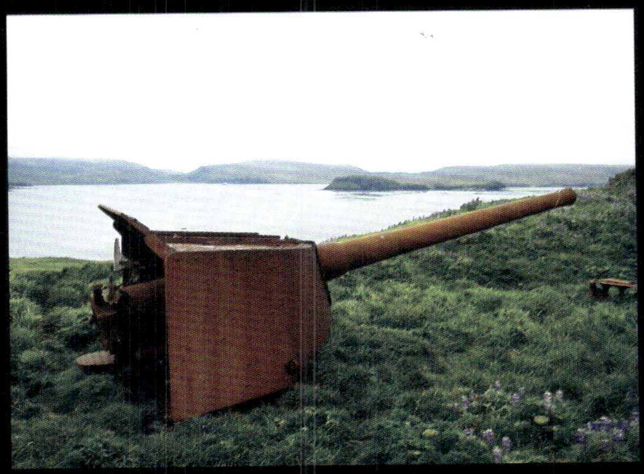

The slow erosion process on shipwrecks like this one is visible in a few protected harbors in the Aleutian Islands. Attu and Kiska islands were the only U.S. soil occupied by Japanese forces during World War II. Japanese troops held Kiska Island from June 1942 to August 1943. *Kiska Harbor, Kiska Island, S. Jewett photo.*

Remnants of World War II defense structures dot the coastline of the Aleutians. Gun emplacements like this one are often found where they once protected bays and important military caches. *Little Kiska Island, M. Hoberg photo.*

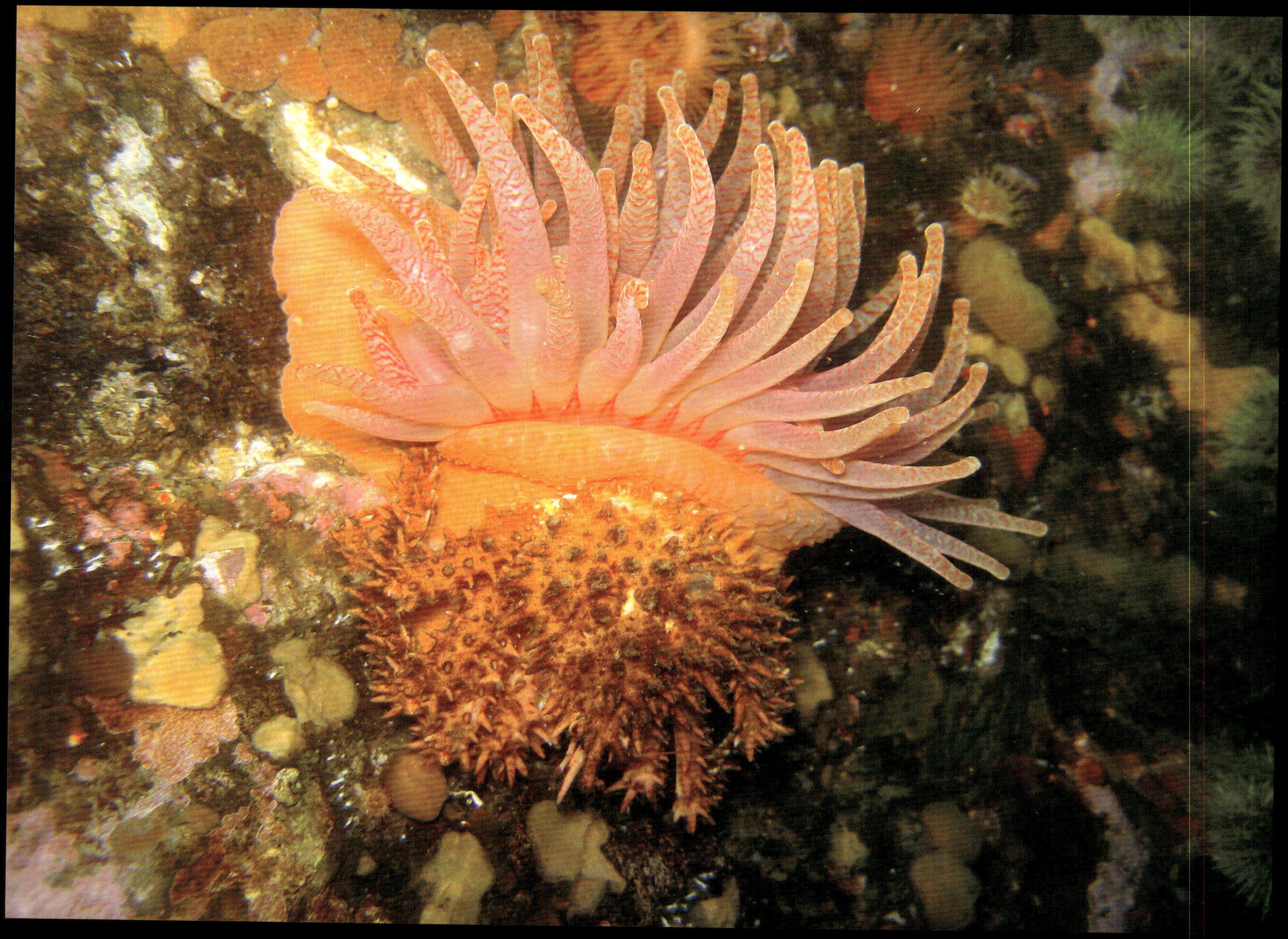

The crimson anemone (*Cribrinopsis fernaldi*) is distinguished from other anemones by the chevron-like shapes on its opaque tentacles. This species often provides shelter to shrimps or crabs—in this photo it is a resting spot for a juvenile king crab. *Rat Island, R. Brewer photo.*

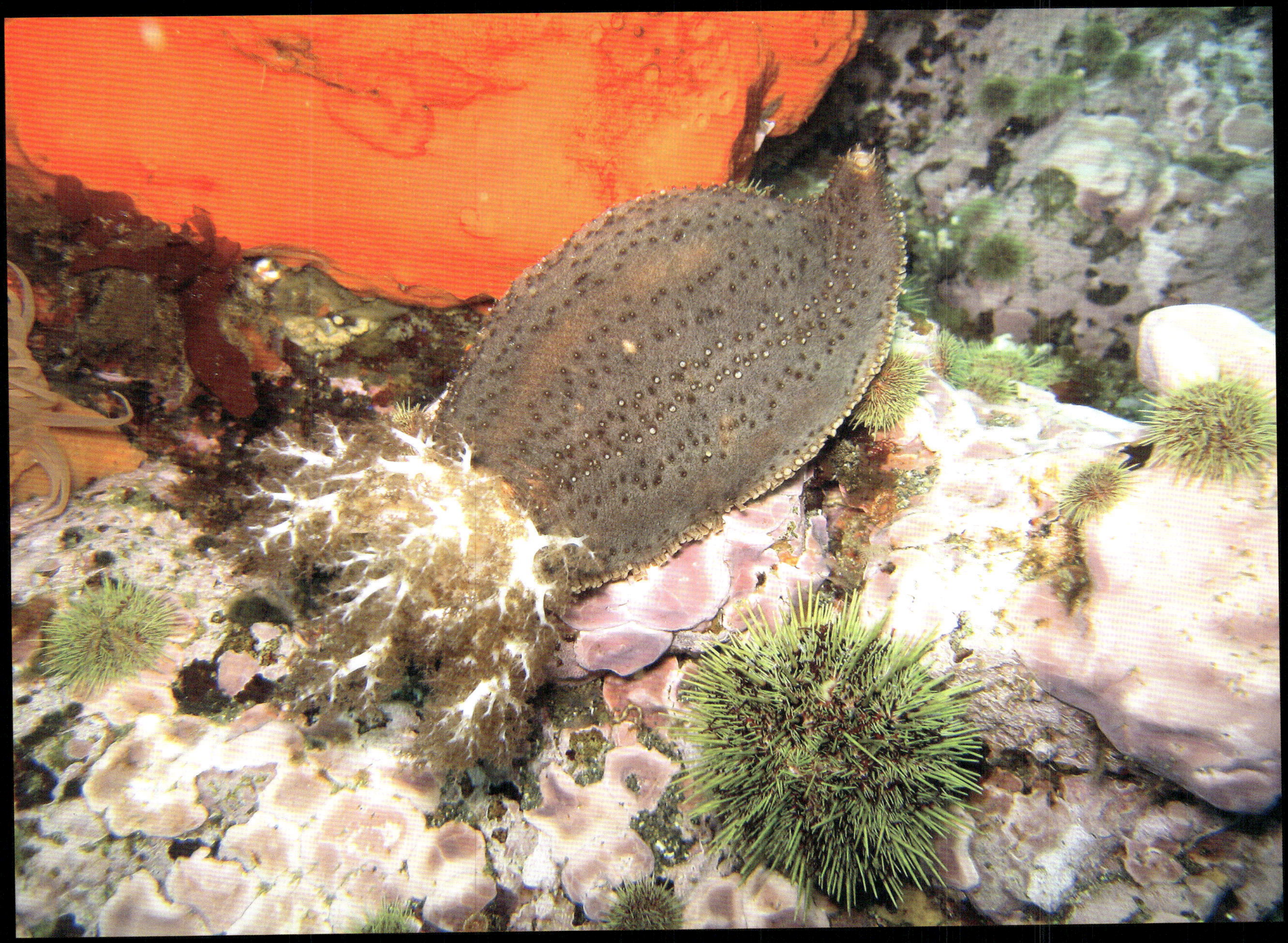

The giant black cucumber (*Cucumaria frondosa japonica*) uses its finely branched tentacles to filter-feed by catching particles in the water. Sea cucumbers are relatives of sea stars, sea urchins, and sand dollars. This species is found throughout the Aleutians, often wedging its extendable body in boulder fields. *Haycock Rock, Kiska Island, R. Brewer photo.*

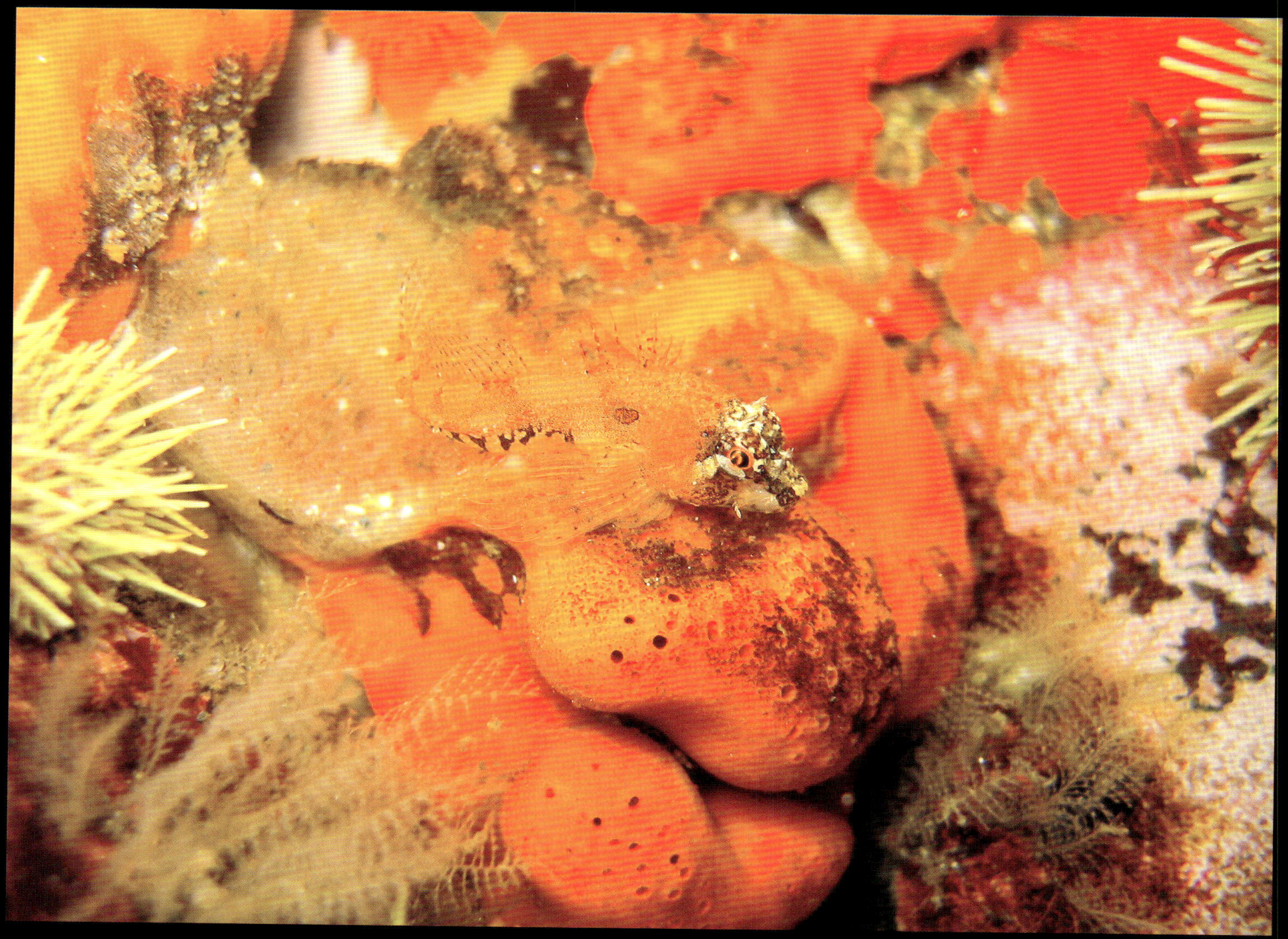

Unidentified common tiny sculpins (Cottidae) demonstrate mimicry, matching perfectly with their backgrounds. Sculpins usually have large heads and round bodies with fanlike pectoral fins. Most of the 300 species of sculpins are found in temperate marine and freshwater habitats of the Northern Hemisphere. *Amchitka Island, H. Chenelot photos.*

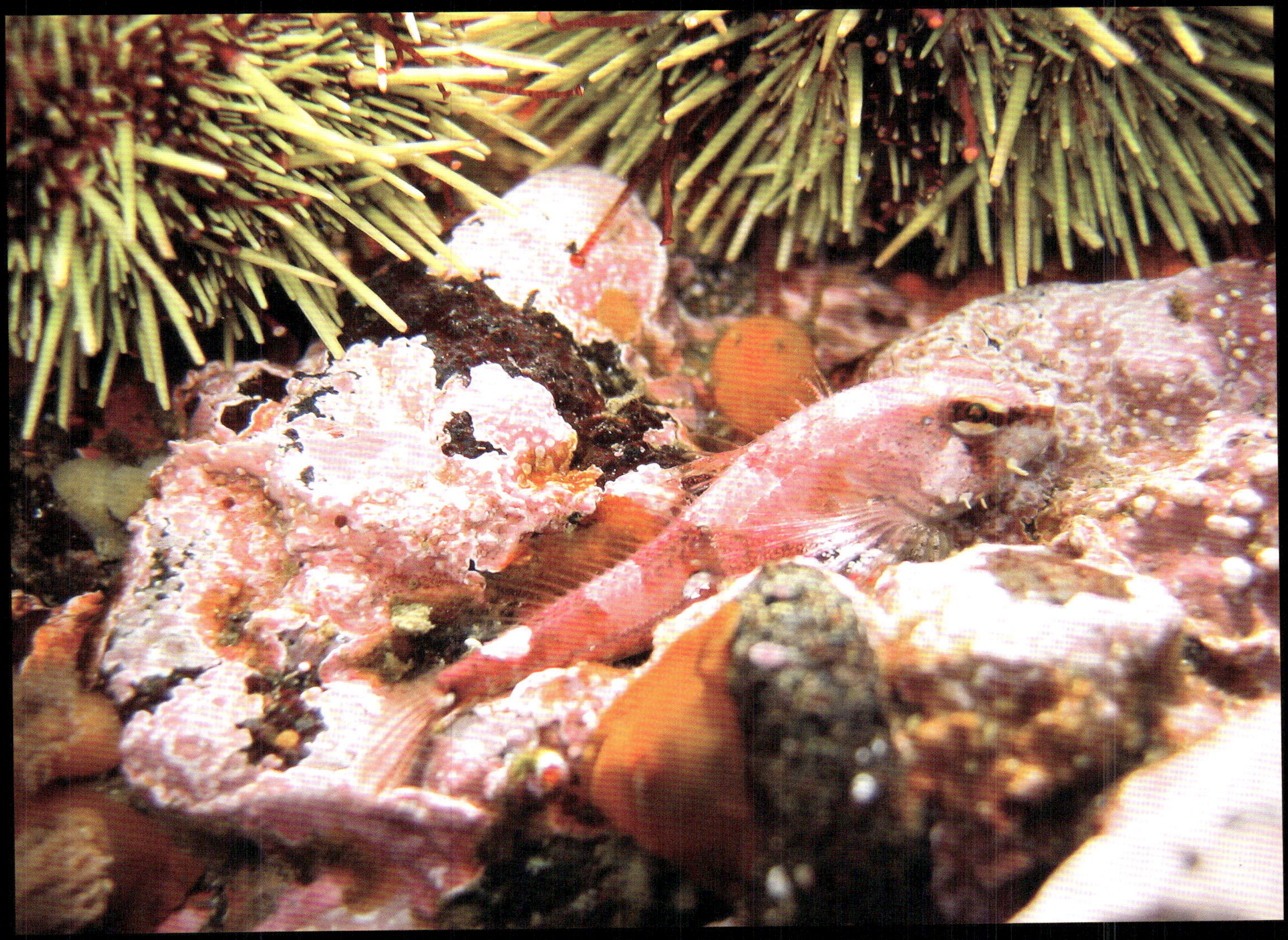

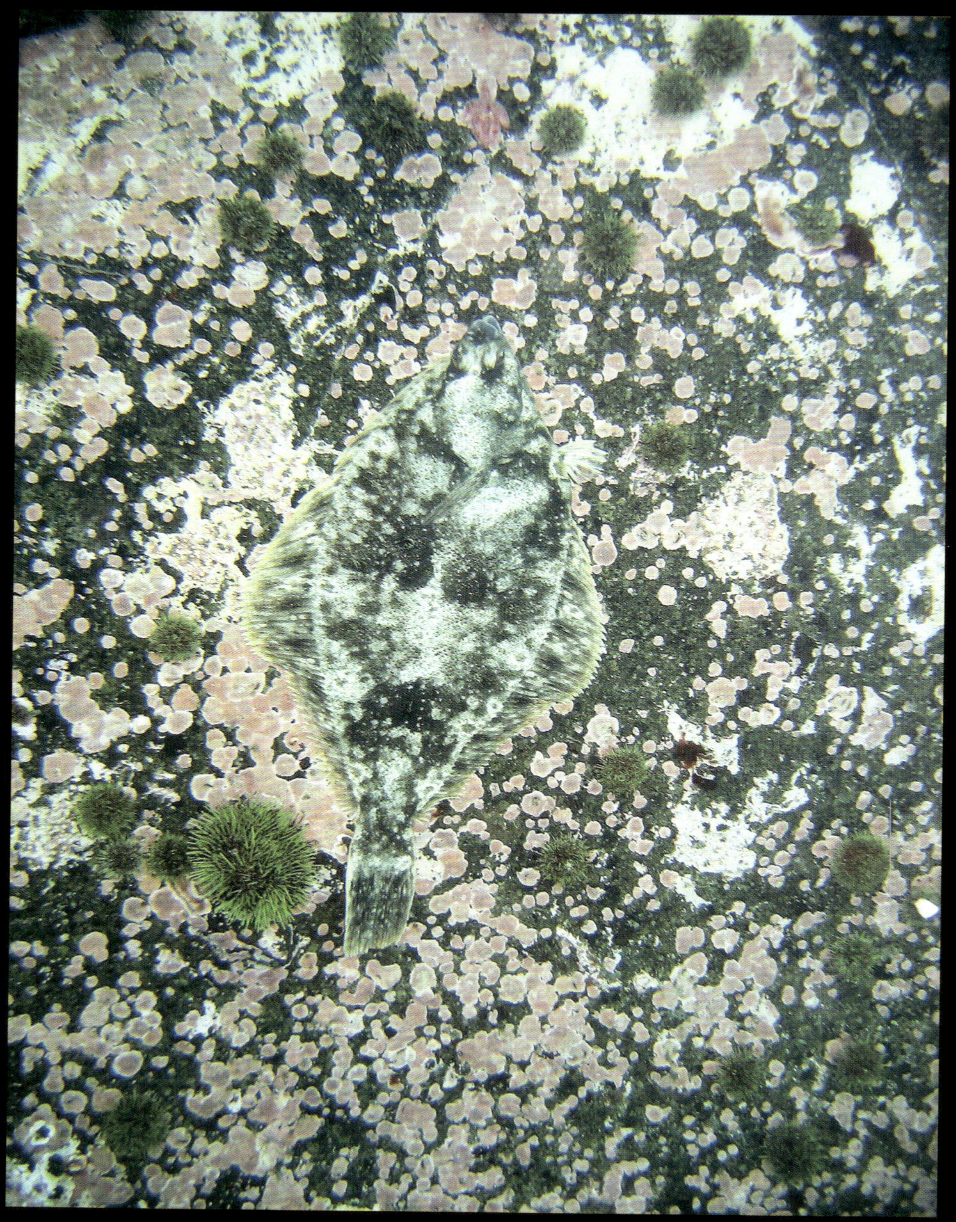

The northern rock sole (*Lepidopsetta polyxystra*) is found from the northern Bering Sea, throughout the Aleutian Islands, to Puget Sound. The rock sole grows to 24 in (60 cm) and can live beyond 20 years. This fish lives on sand, mud, gravel, and other surfaces where it is well camouflaged and eats small invertebrates. *Rat Island, H. Chenelot photos.*

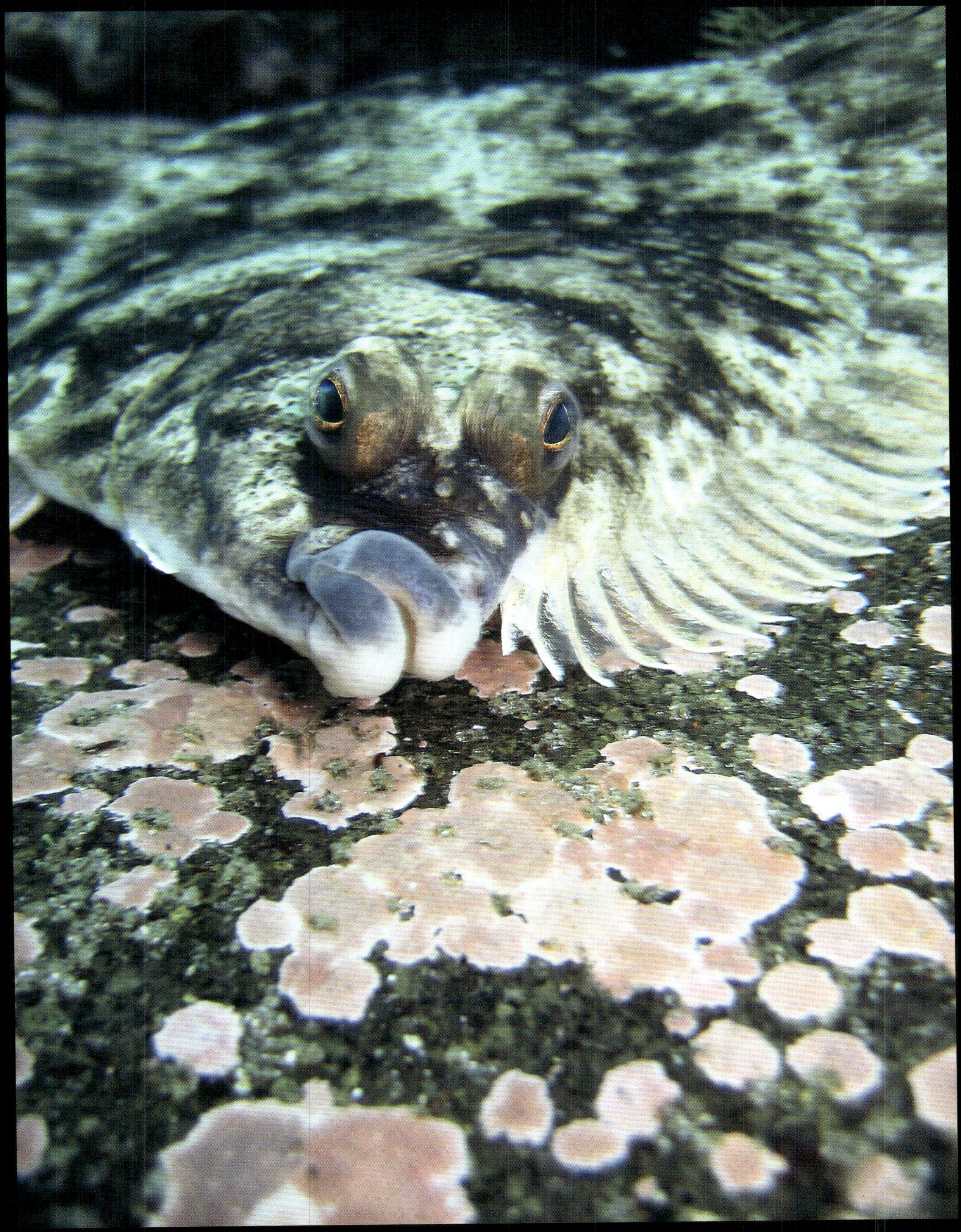

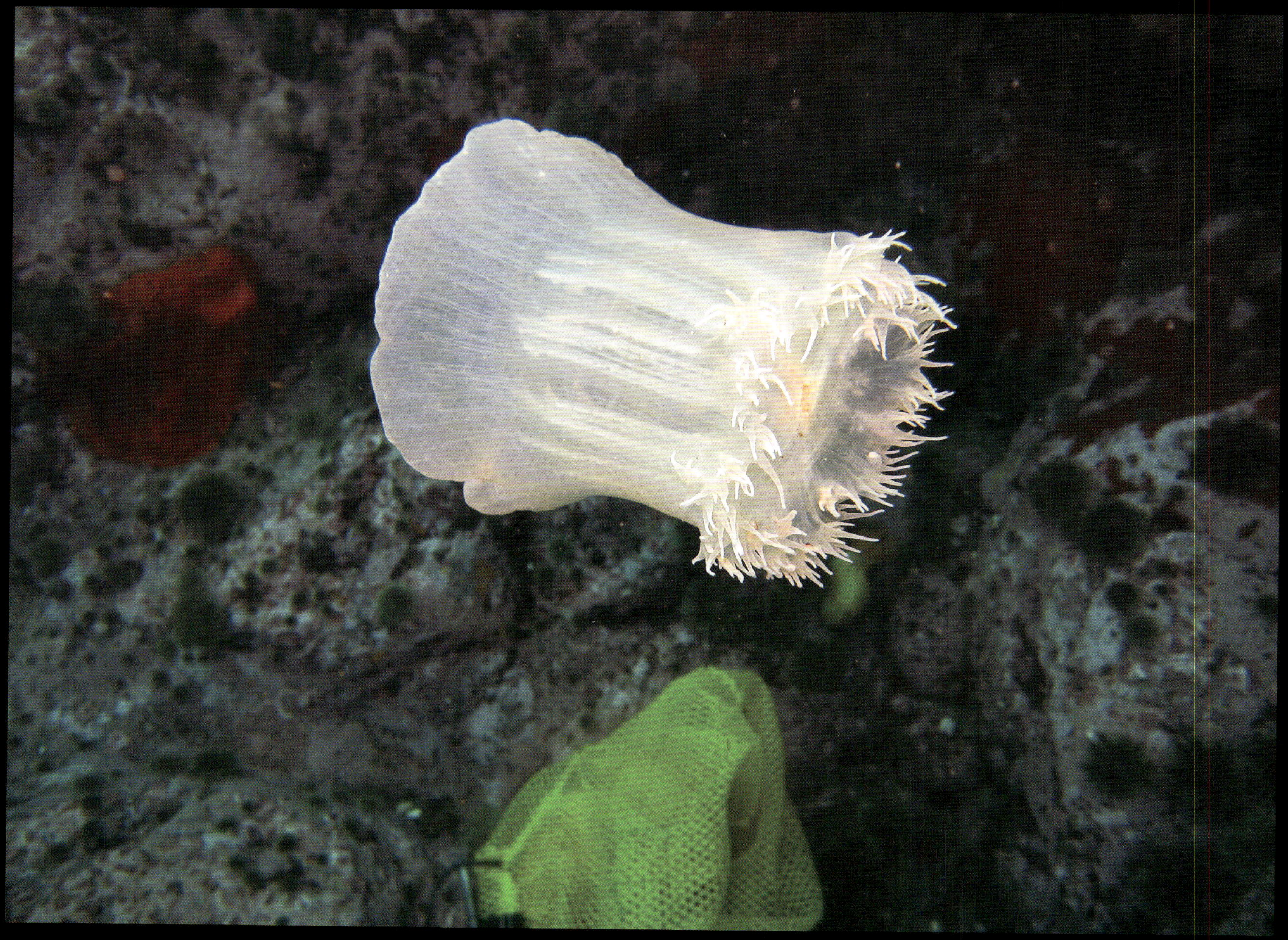

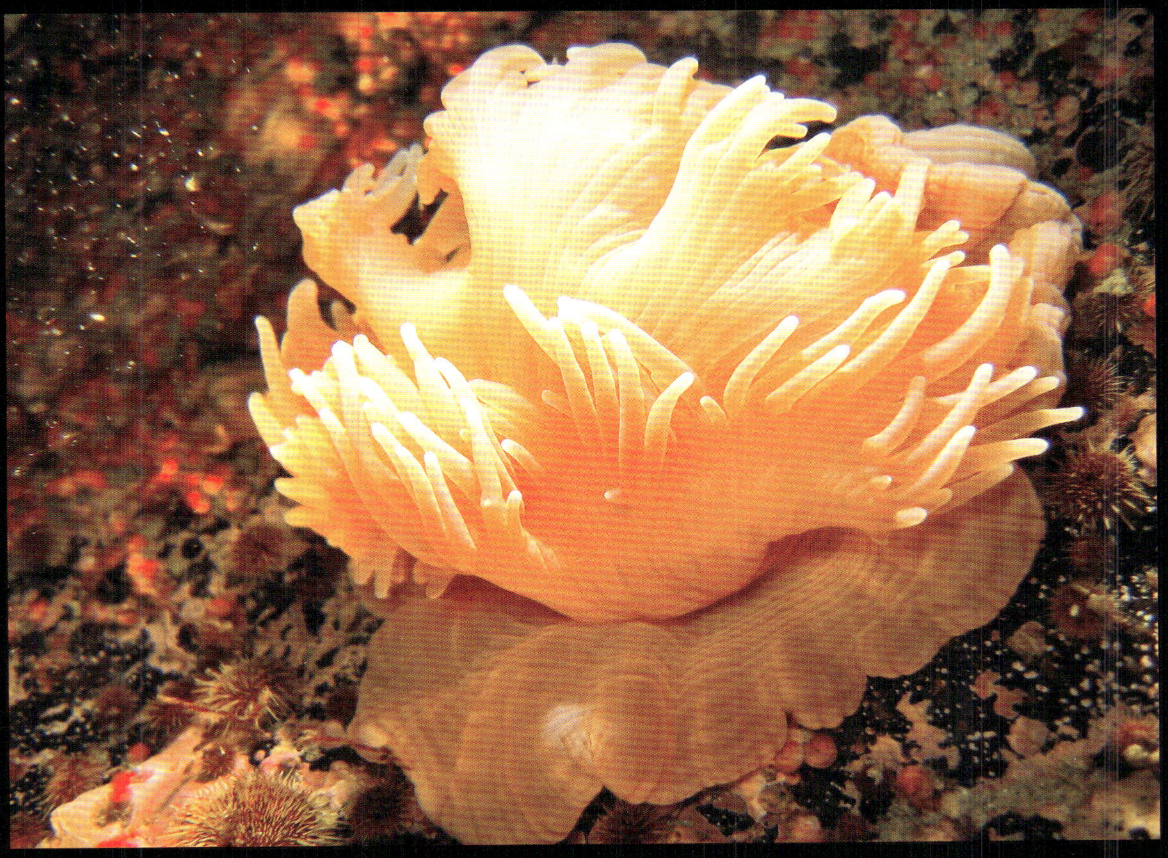

The swimming anemone (*Ptychodactis aleutiensis*) is aptly named for its ability to dislodge and be carried by the current. Its shape constantly changes. Although it occurs in all five Aleutian Island groups, it is more common in the western and central Aleutians. *Ptychodactis aleutiensis* was recently described and named. *In motion, Haycock Rock, Kiska Island. Stationary, Kirilof Point, Amchitka Island, H. Chenelot photos.*

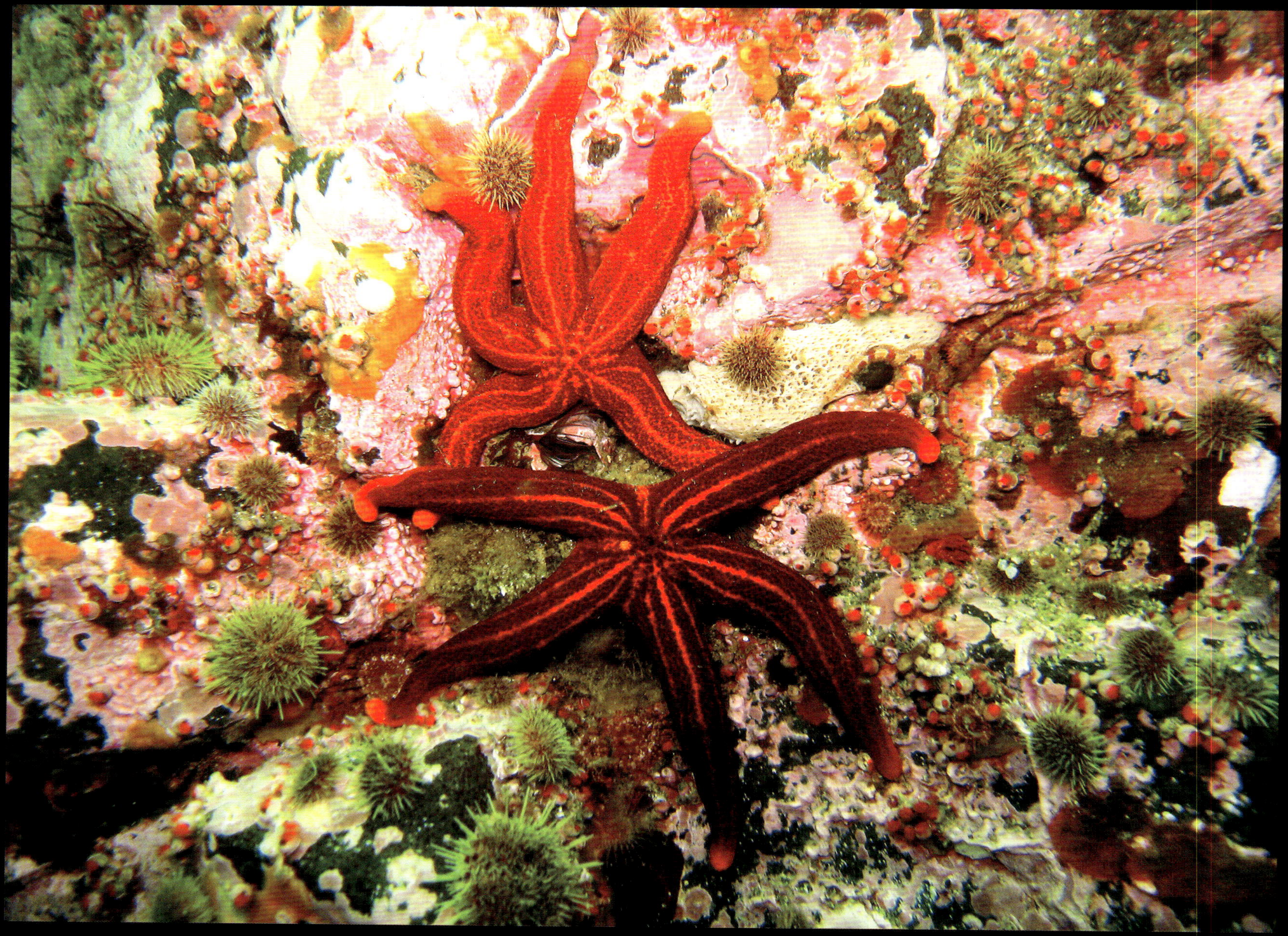

Striped henricia (*Henricia lineata*) has narrow, elongate arms, a small central disc, short spines, and virtually no pedicellariae (pinchers on its back). The striped henricia feeds on plankton, organic matter, bryozoans, and encrusting sponges. This is a recently described new species. *Kirilof Bay, Amchitka Island, S. Jewett photo.*

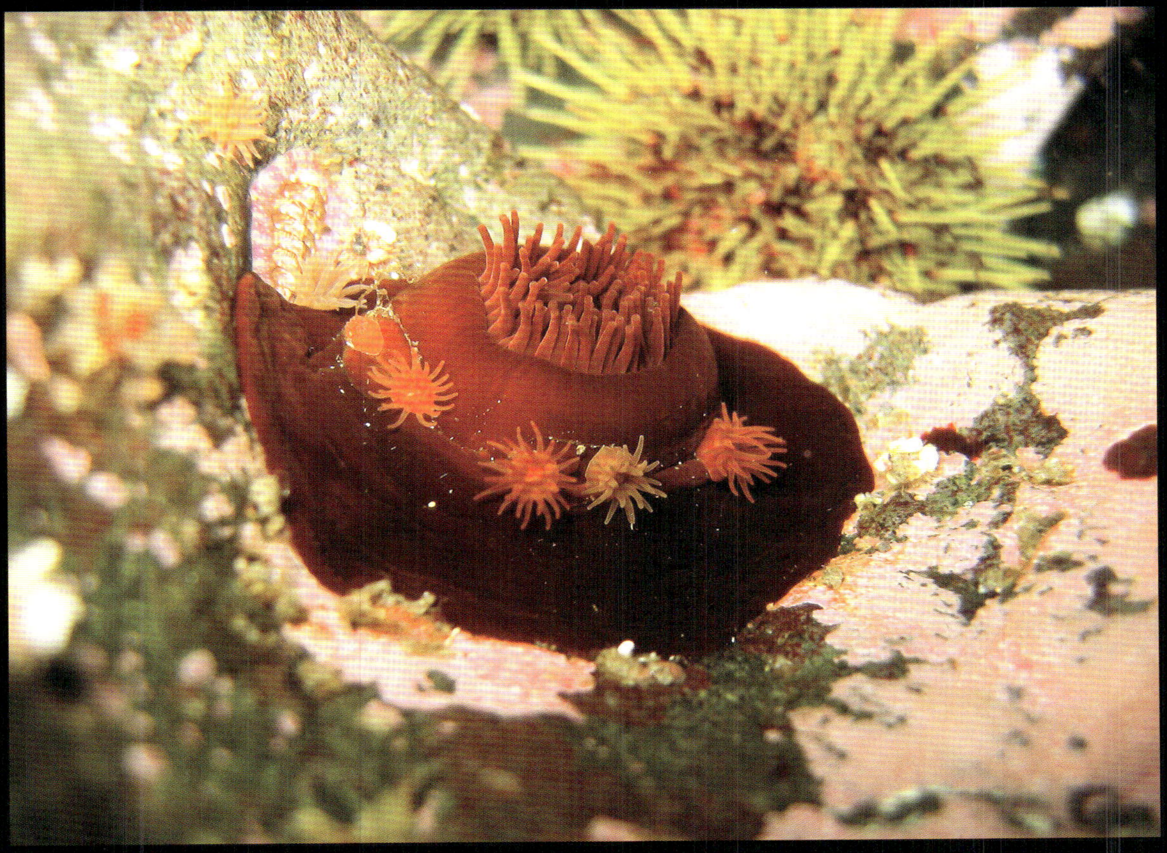

This adult brooding anemone (*Epiactis prolifera*) has juveniles temporarily attached. During sexual reproduction, the eggs are fertilized in the digestive cavity of the adult and the larvae later migrate to the pedal disc where they settle and grow into tiny sea anemones. The species is variable in color and is common throughout the Aleutian Islands to Friday Harbor, Washington. Anemones are relatives of jellyfish, corals, sea pens, and sea whips. *Amchitka Island, H. Chenelot photo.*

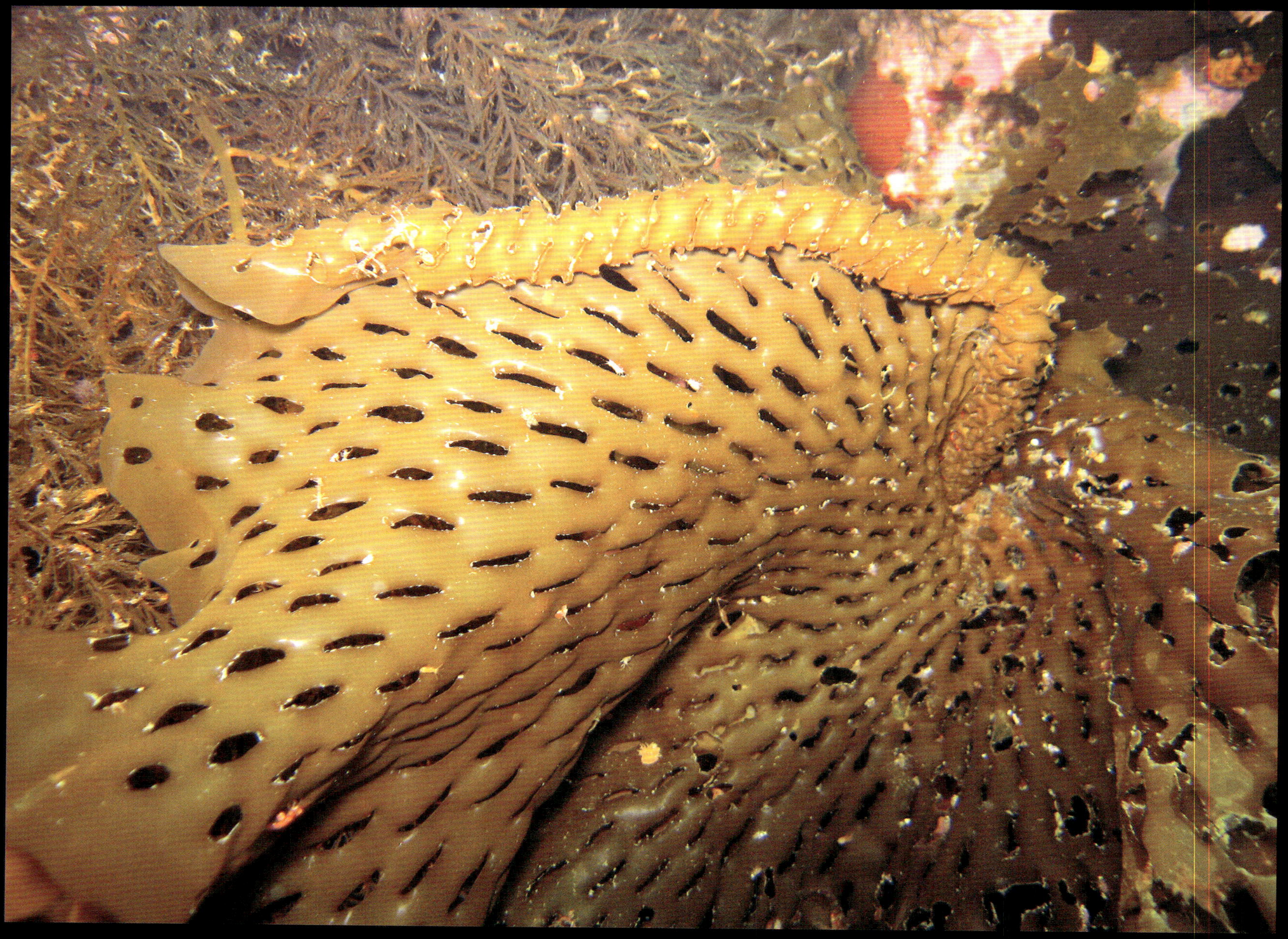

Spiral sieve kelp (*Thalassiophylum clathrus*) is a distinctive member of the nearshore kelp community in the Aleutians. The blade can grow to at least 20 in (50 cm) long with somewhat regular perforations. This perennial kelp is found in low intertidal and shallow subtidal regions that are semi-exposed. *Thalassiophylum clathrus* occurs only in the Pribilof and Aleutian islands. *Rat Island, S. Jewett photo.*

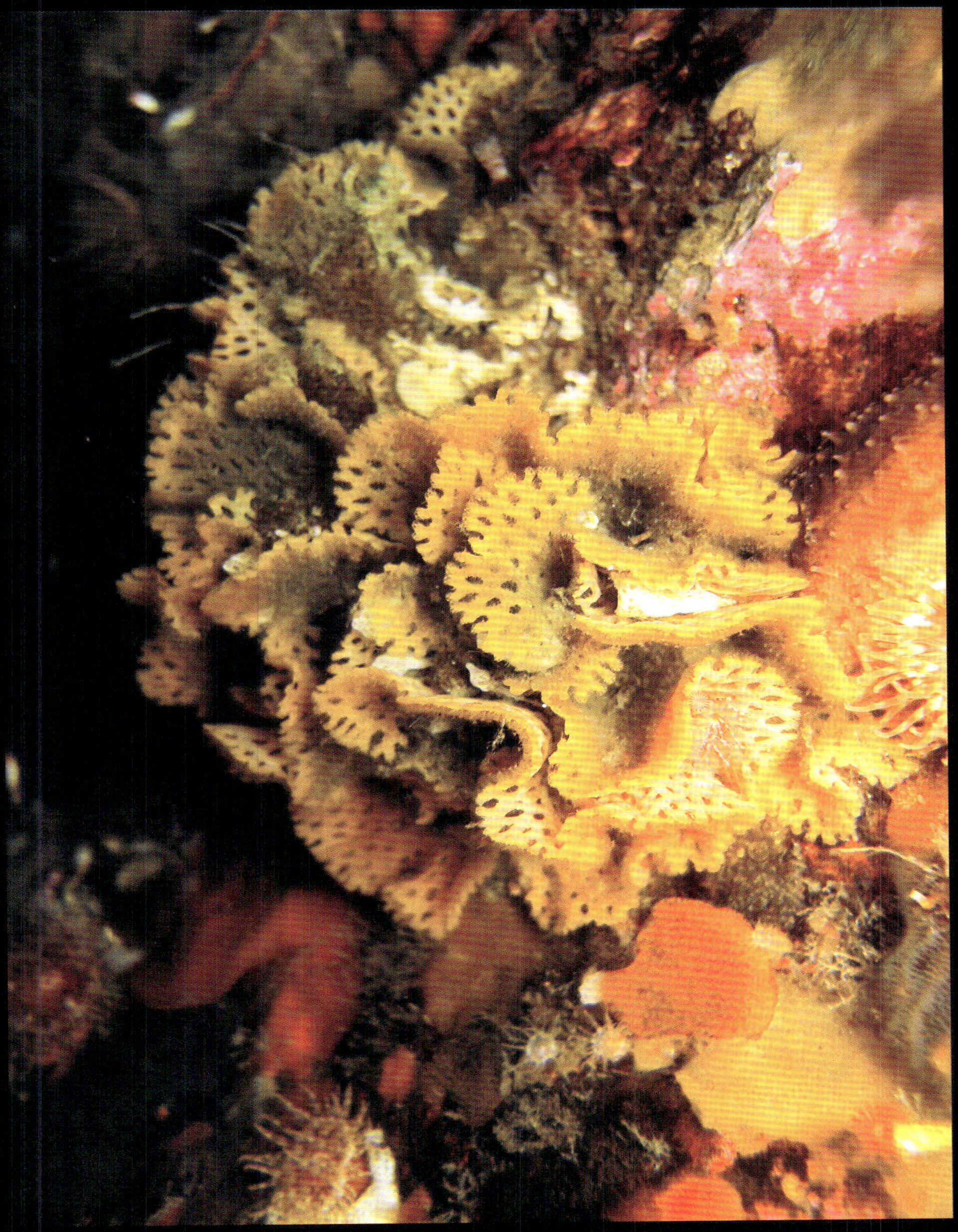

The lattice-work bryozoan (*Phidolopora pacifica*) forms a maze of hiding places for tiny invertebrates and fish. This bryozoan is a delicate colony, whose older central portions are white and lifeless. Bryozoans like this one are preyed upon by sea stars, snails, and nudibranchs. *Finger Point, Little Sitkin Island, R. Brewer photo.*

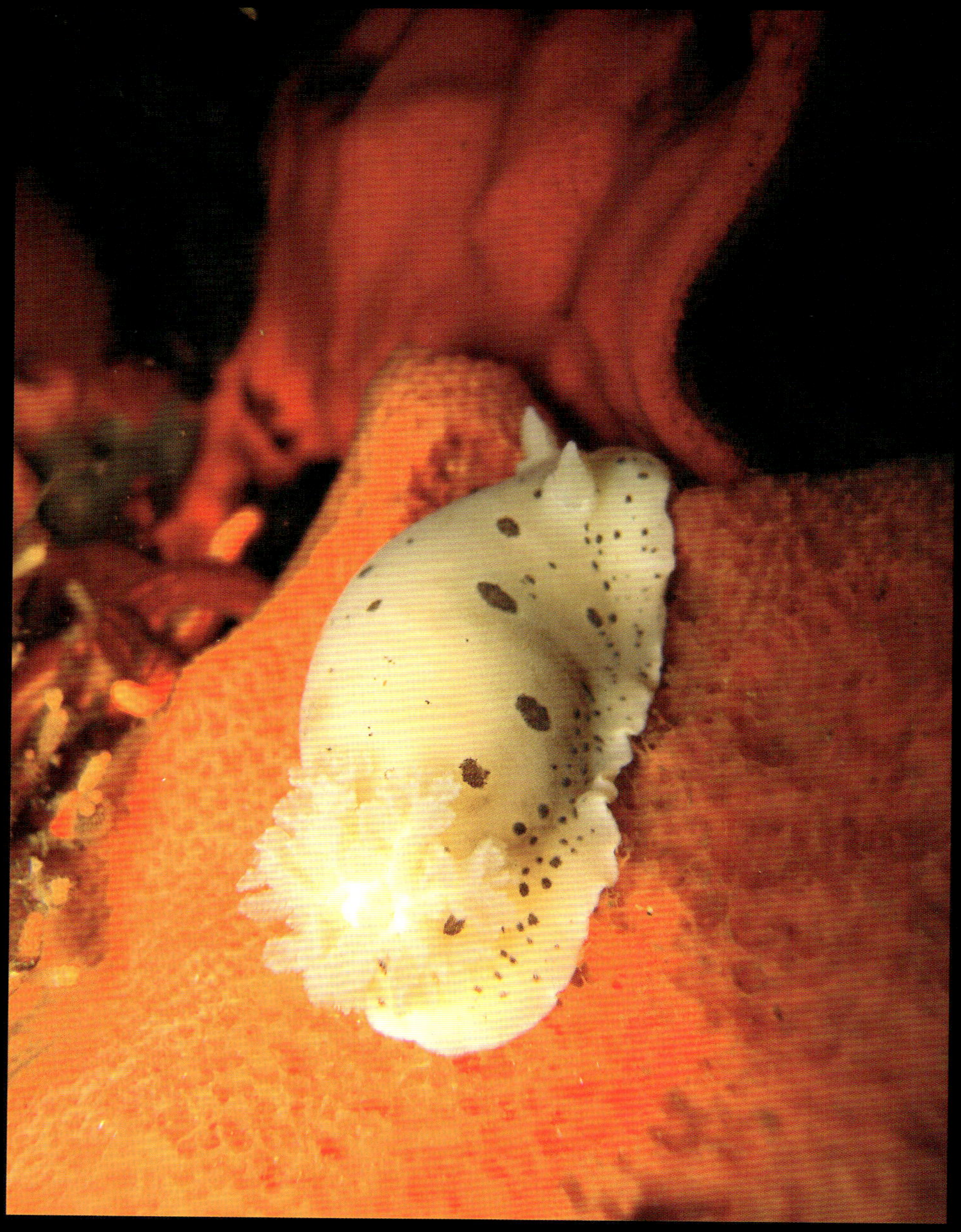

The leopard dorid (*Diaulula sandiegensis*) is distinguished by the dark rounded spots covering its white to brown body. Nudibranchs are relatives of snails but most lack a shell. The leopard dorid feeds primarily on sponges and is found throughout the Aleutians. *Rat Island, R. Brewer photo.*

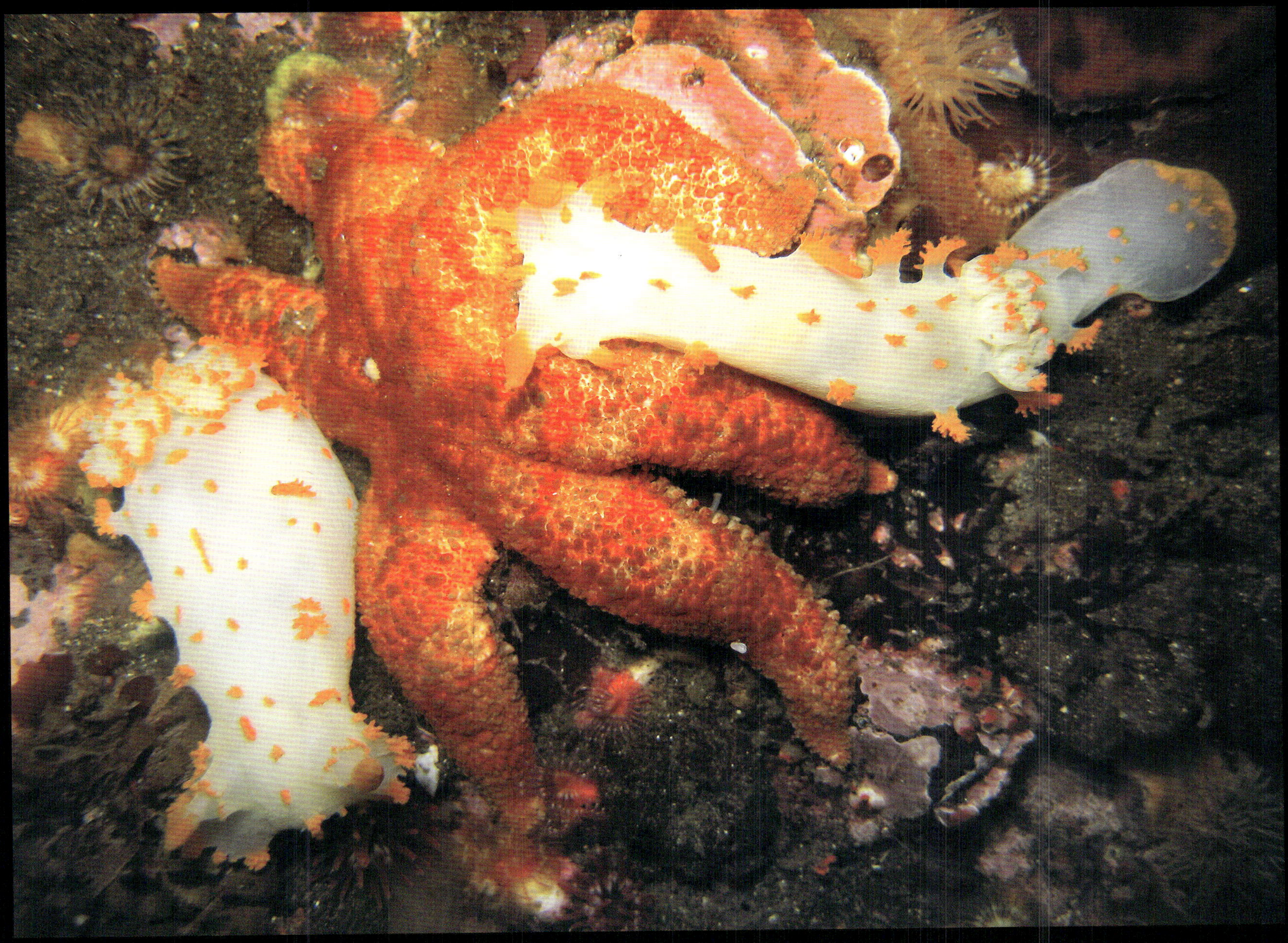

Orange tip or clown nudibranchs (*Triopha catalinae*) feed on bryozoans and are found all along the Aleutian Islands. These nudibranchs are usually white with 8 to 12 irregular orange processes. In the photo two nudibranchs "embrace" a six-armed sea star (*Leptasterias* sp.). *Amchitka Island, R. Brewer photo.*

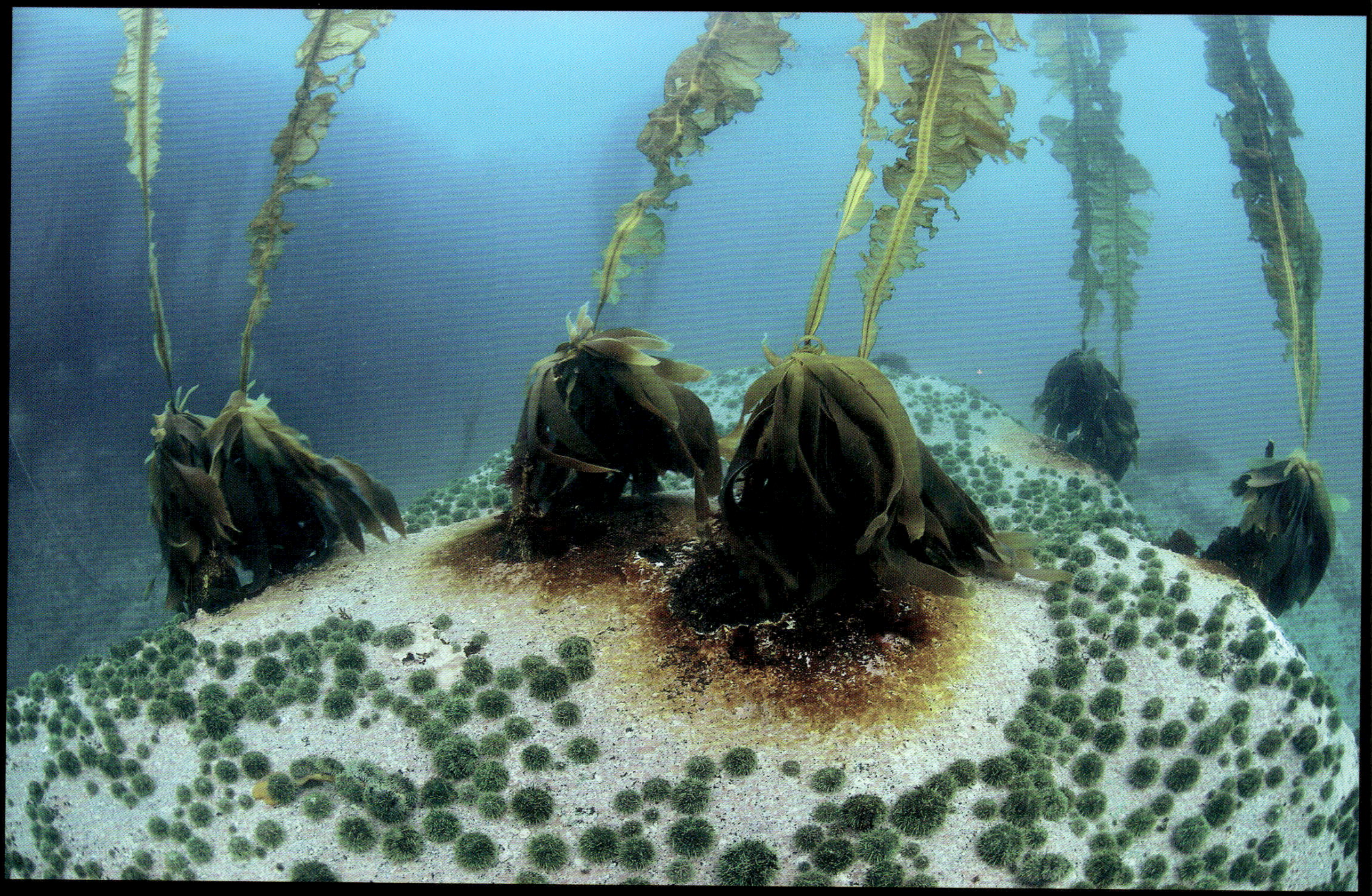

Like sweeping brooms, leaflike sporophylls of the dragon kelp (*Eualaria fistulosa*) keep would-be grazing sea urchins away. *Finger Point, Little Sitkin Island, S. Harper photo.*

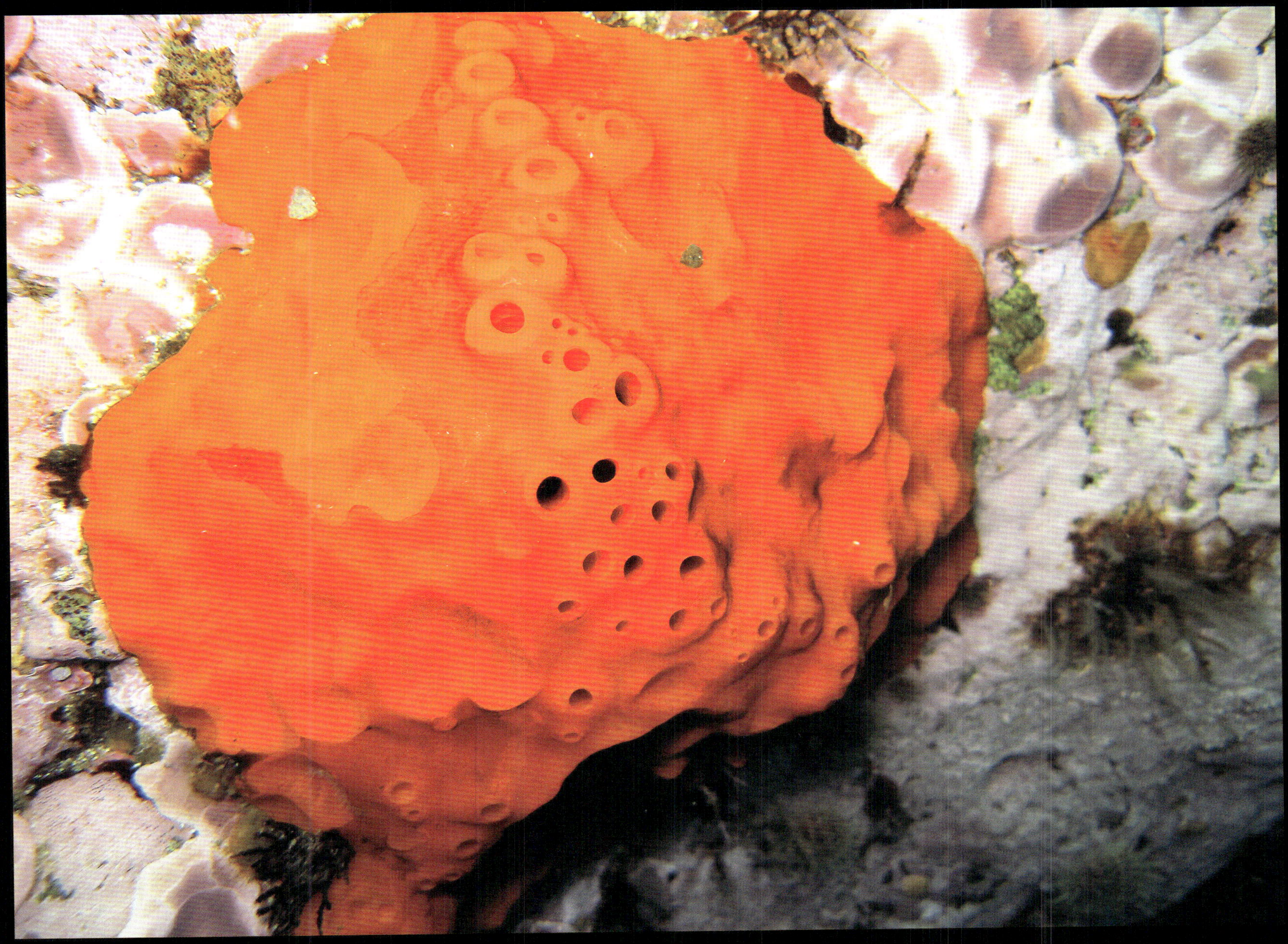

The prominent oscula (pores) of the orange sponge are the outlets for filtered seawater. This species will grow to a height of about 6 in (15 cm) and can be found on rocks throughout the Aleutians. *Haycock Rock, Kiska Island, R. Brewer photo.*

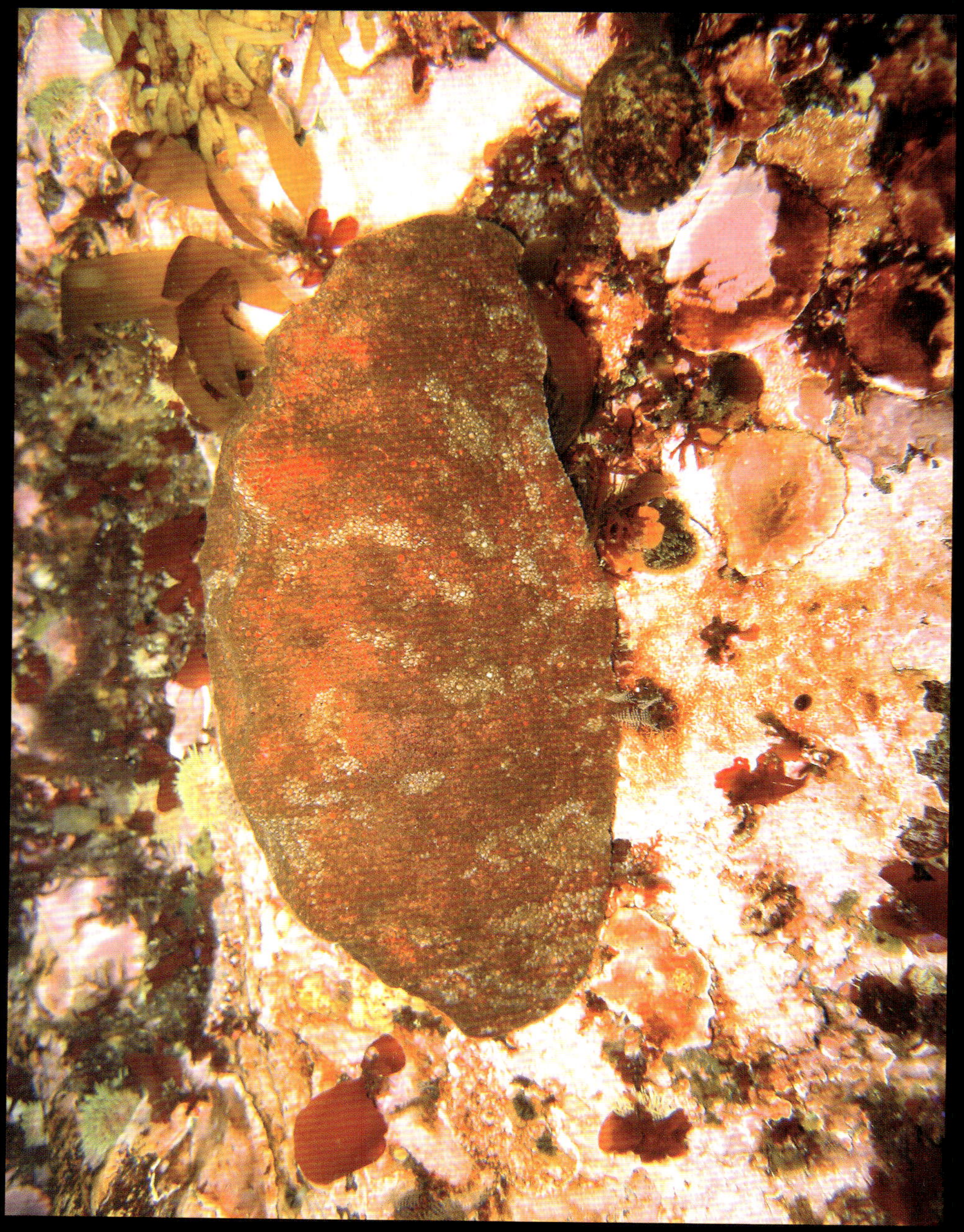

The gumboot or Steller's chiton (*Cryptochiton stelleri*) is the largest chiton in the North Pacific, attaining a maximum length of 14 in (35 cm). Unlike other chiton species, the eight valves (shells) of the gumboot chiton are hidden. It can live to 20 years and eats red algae. Native people of the Aleutians collect this species as a food resource, during low tide. *Yug Point, Little Kiska Island, R. Brewer photo.*

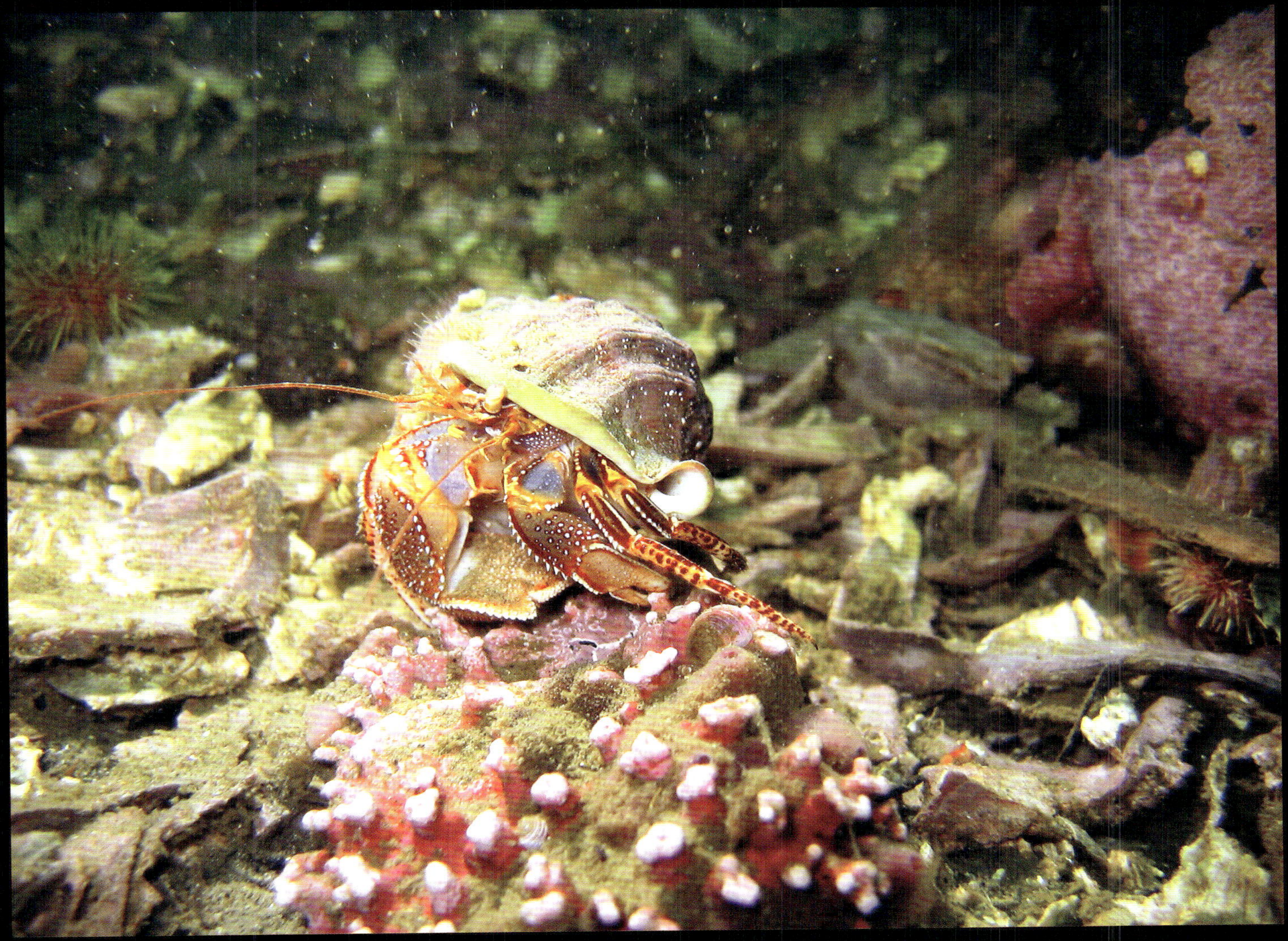

The widehand hermit crab (*Elassochirus tenuimanus*) wanders through various habitat types. It is found in the Bering Sea and throughout the Aleutians to Washington, at depths less than 600 ft (182 m). This species has a distinctively wide right chela (claw) and blue coloration at the base of its legs. It often has tiny hydroids and polychaetes on its shell. *Kiska Island, H. Chenelot photo.*

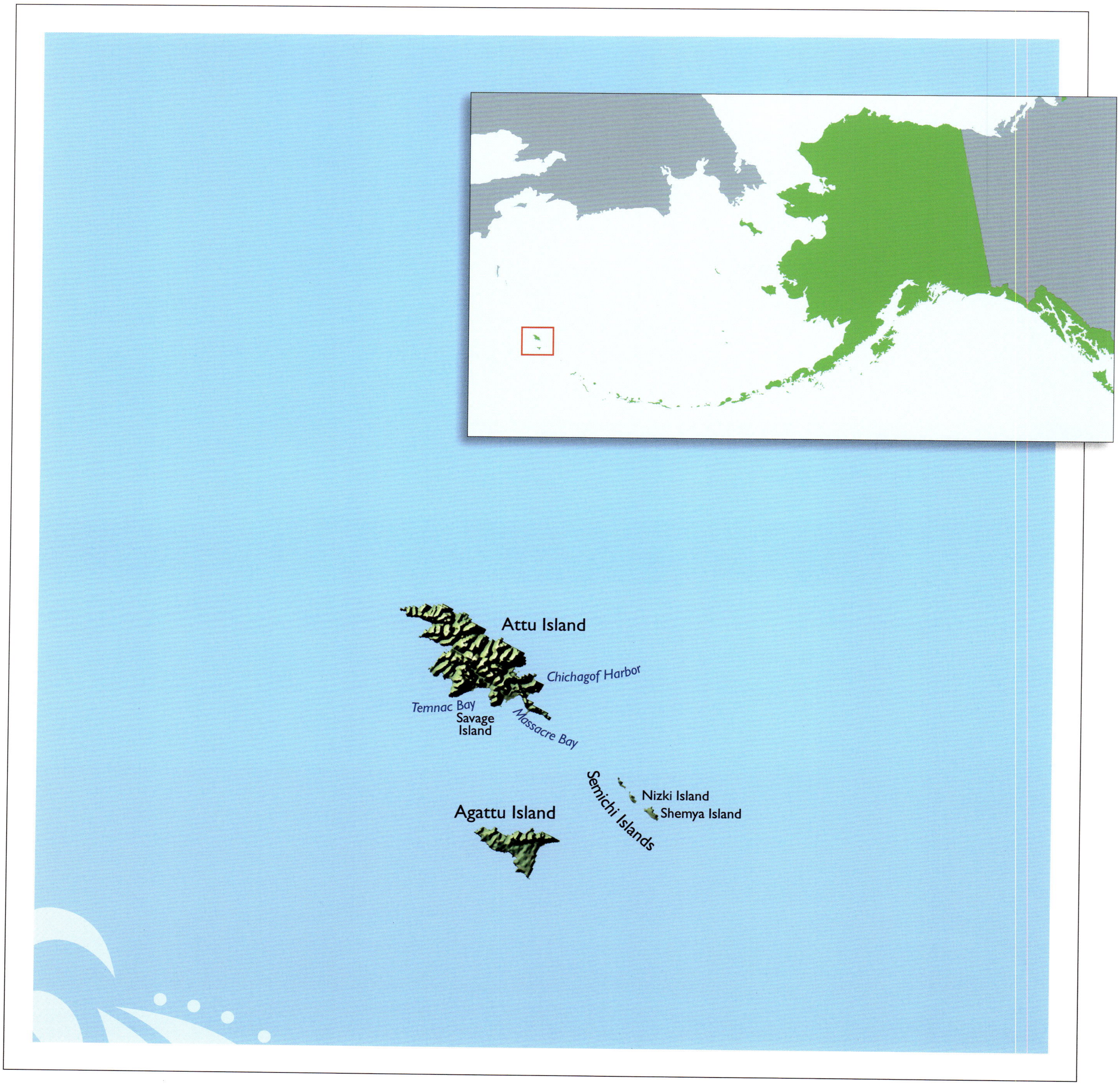
Attu Island
Chichagof Harbor
Temnac Bay
Savage Island
Massacre Bay
Semichi Islands
Nizki Island
Shemya Island
Agattu Island

Near Islands

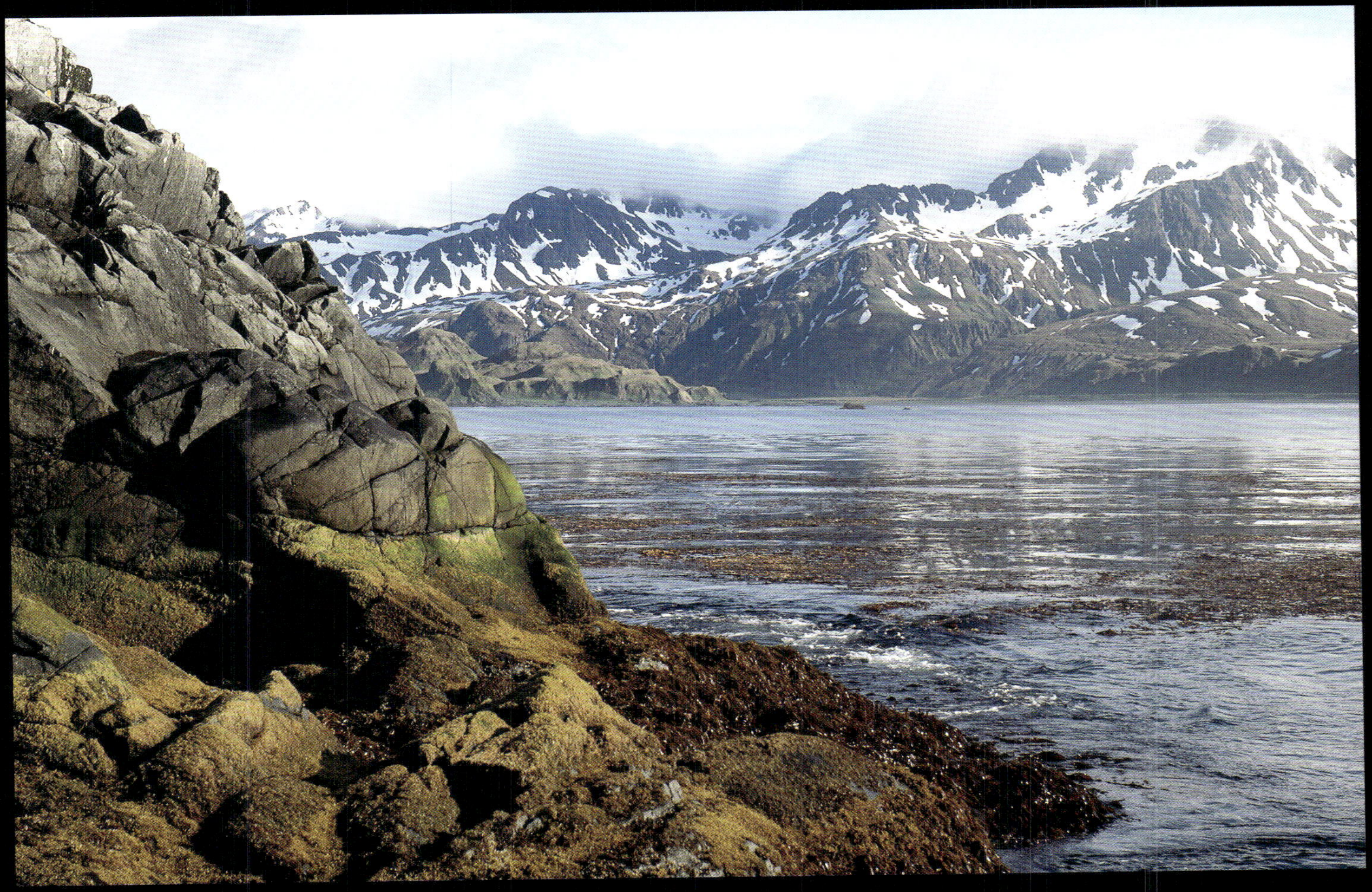

Attu Island is the westernmost and largest of the Near Islands, making it the farthest-west point of land in Alaska and the United States, nearly 1,200 miles from mainland Alaska. Attu was occupied by Japanese forces during World War II. *Temnac Bay, Attu Island, M. Lindeberg photo.*

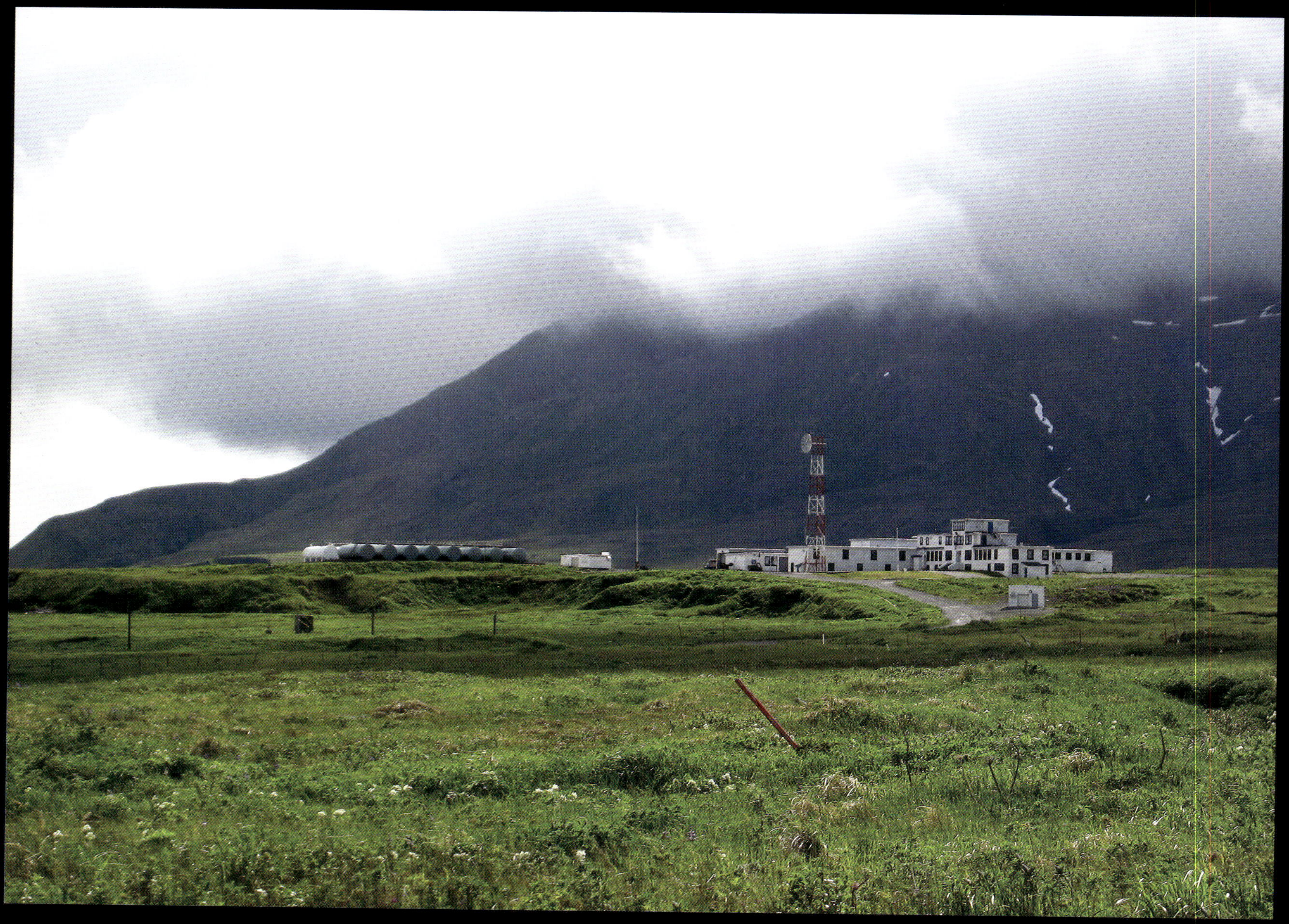

This Loran station was manned by 20 Coast Guard personnel until it closed in late 2010, ending inland habitation. Prior to the Japanese invasion of Attu during World War II, up to 50 Aleut Natives lived there. *Massacre Bay, Attu Island, N. Dallman photo.*

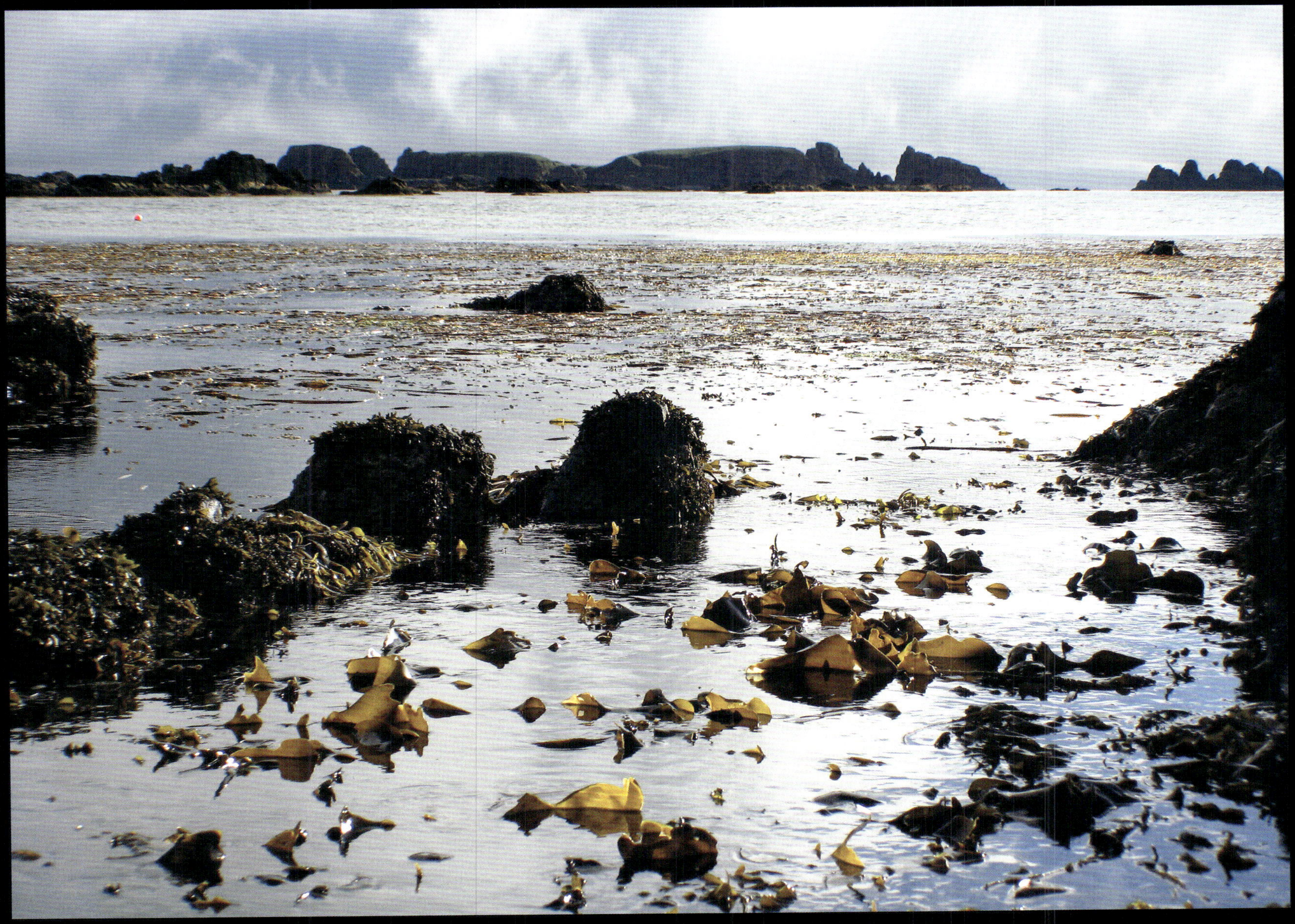

Low tide reveals thick-bladed brown kelp near Attu Island. These kelp forests provide refuge, habitat, and food for marine invertebrates and fishes. *Chichagof Harbor, Attu Island, M. Lindeberg photo.*

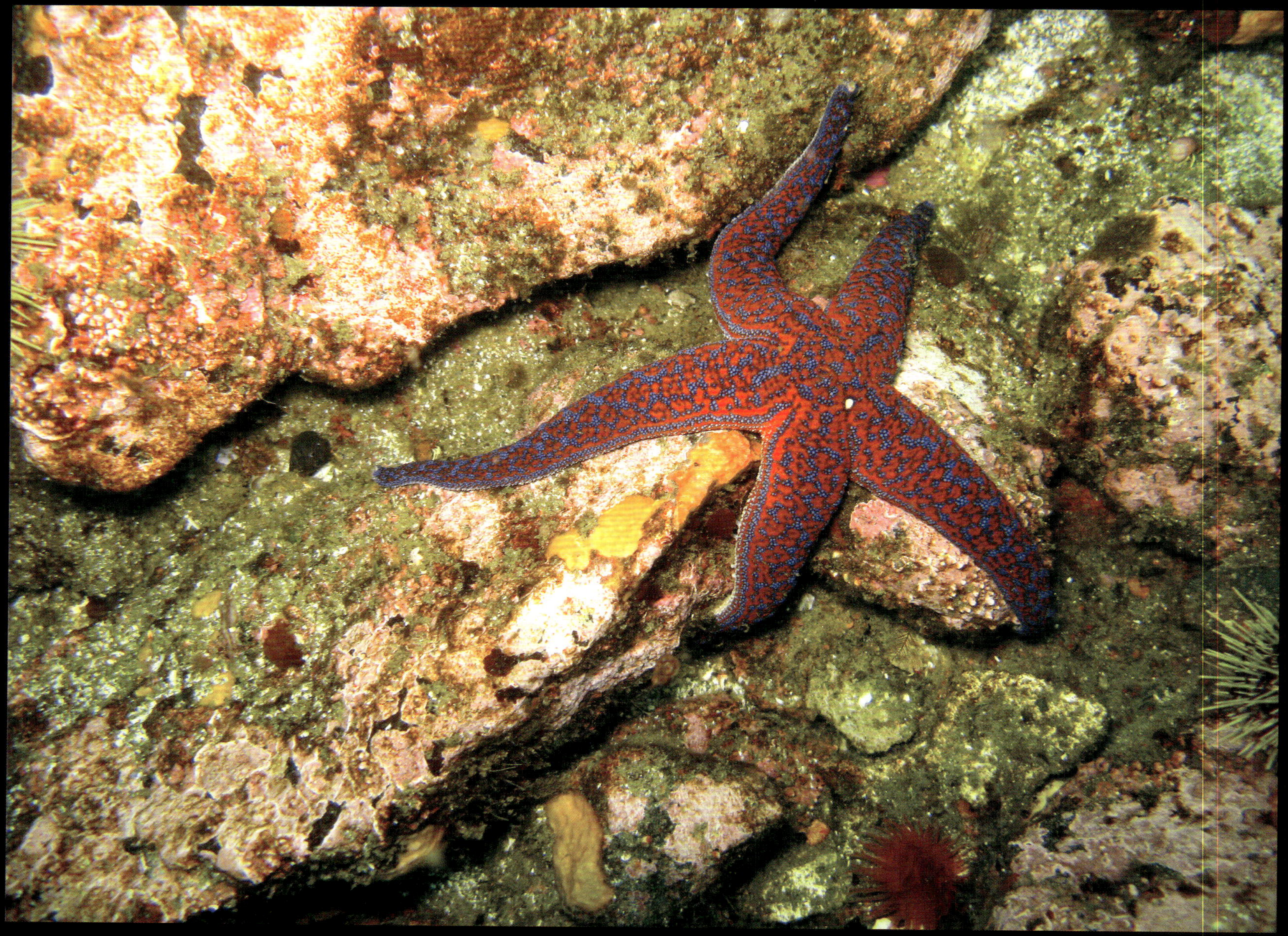

The blue-spined sea star (*Evasterias retifera*) is large (to 16 in, 40 cm diameter) and colorful. It occurs in the western Aleutians, west of Buldir Pass, in the Near Islands to Kamchatka and to the Kuril Islands. This species can be confused with *E. troschelii* where the two species co-occur, but *E. retifera* can be distinguished by its brilliant coloration—red to red-orange with blue spines. *Massacre Bay, Attu Island, H. Chenelot photo.*

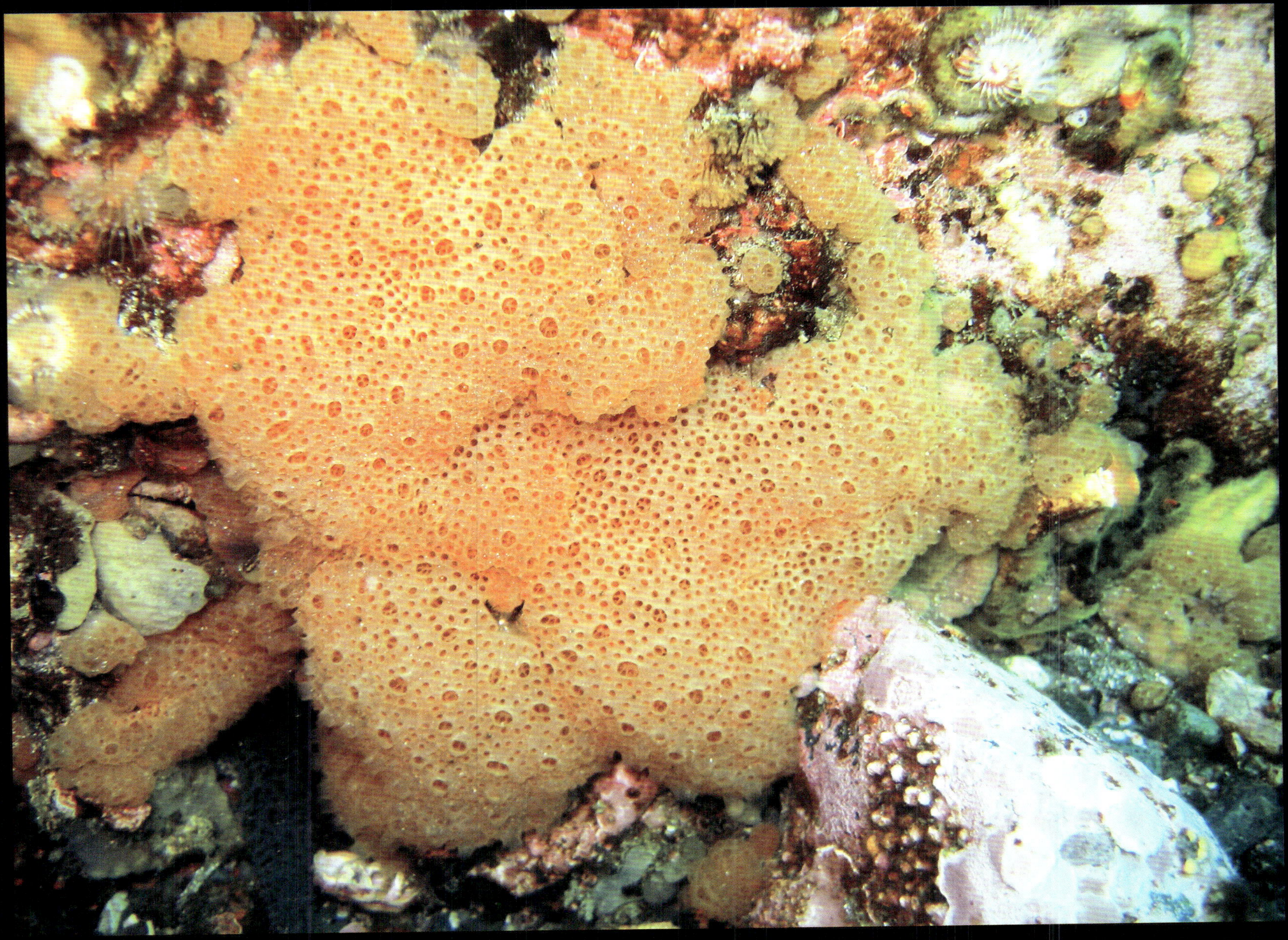

Ascidians are a diverse group of organisms, with many species still unidentified. Ascidians are filter-feeders that exist in three forms: solitary, social, and compound (pictured here). There is concern that non-native ascidians may be invading Alaska waters and outcompeting other attached organisms for space. *Attu Island, H. Chenelot photo.*

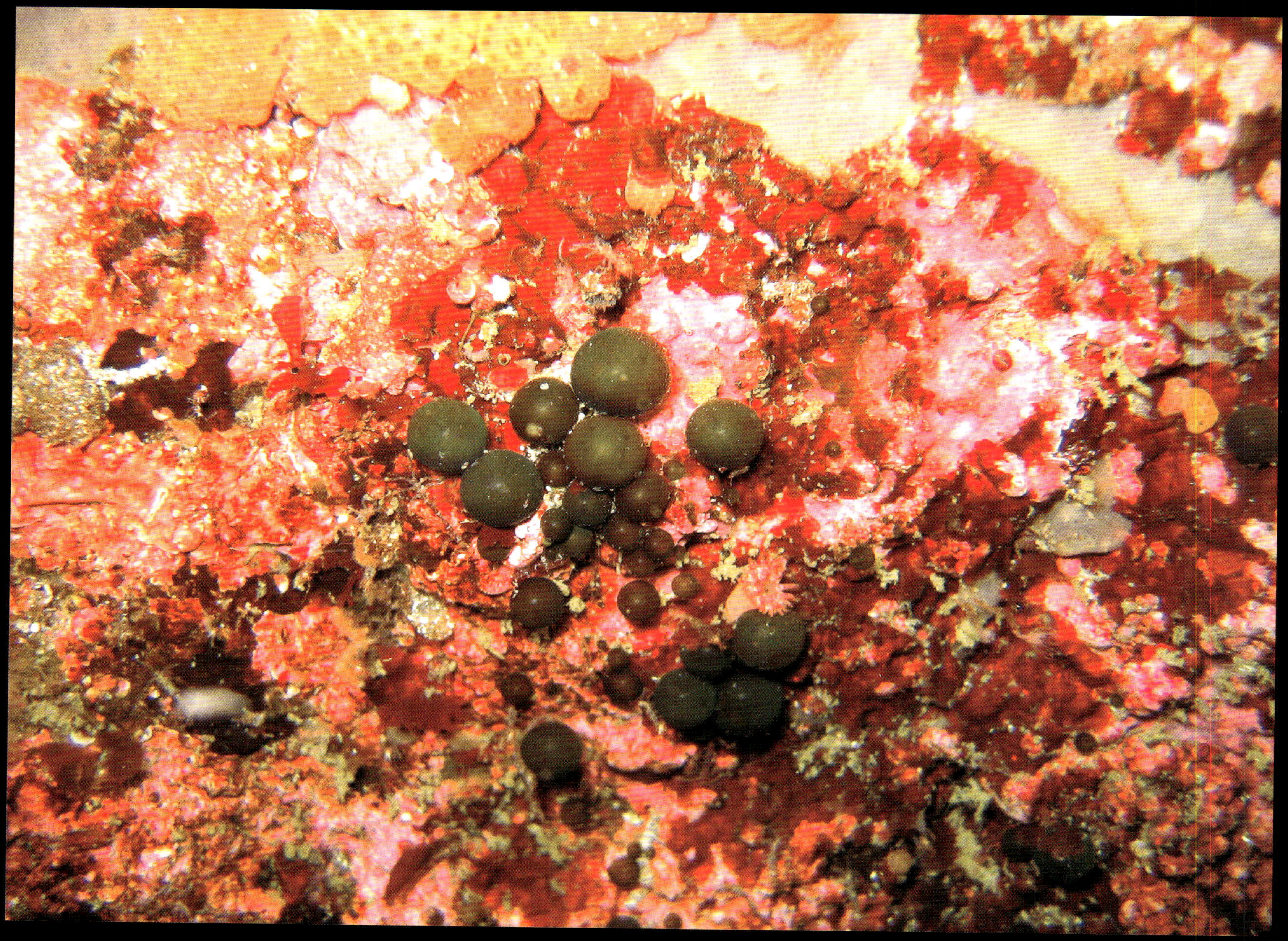

The green sea grape (*Derbesia marina*), also known as the sea pearl, has two distinct life cycle phases—the single-celled gamete shown here (0.1-0.6 in, 2-15 mm wide) and the filamentous, spore-producing phase. This green alga is found from northern Alaska to Mexico. *Shemya Island, H. Chenelot photo.*

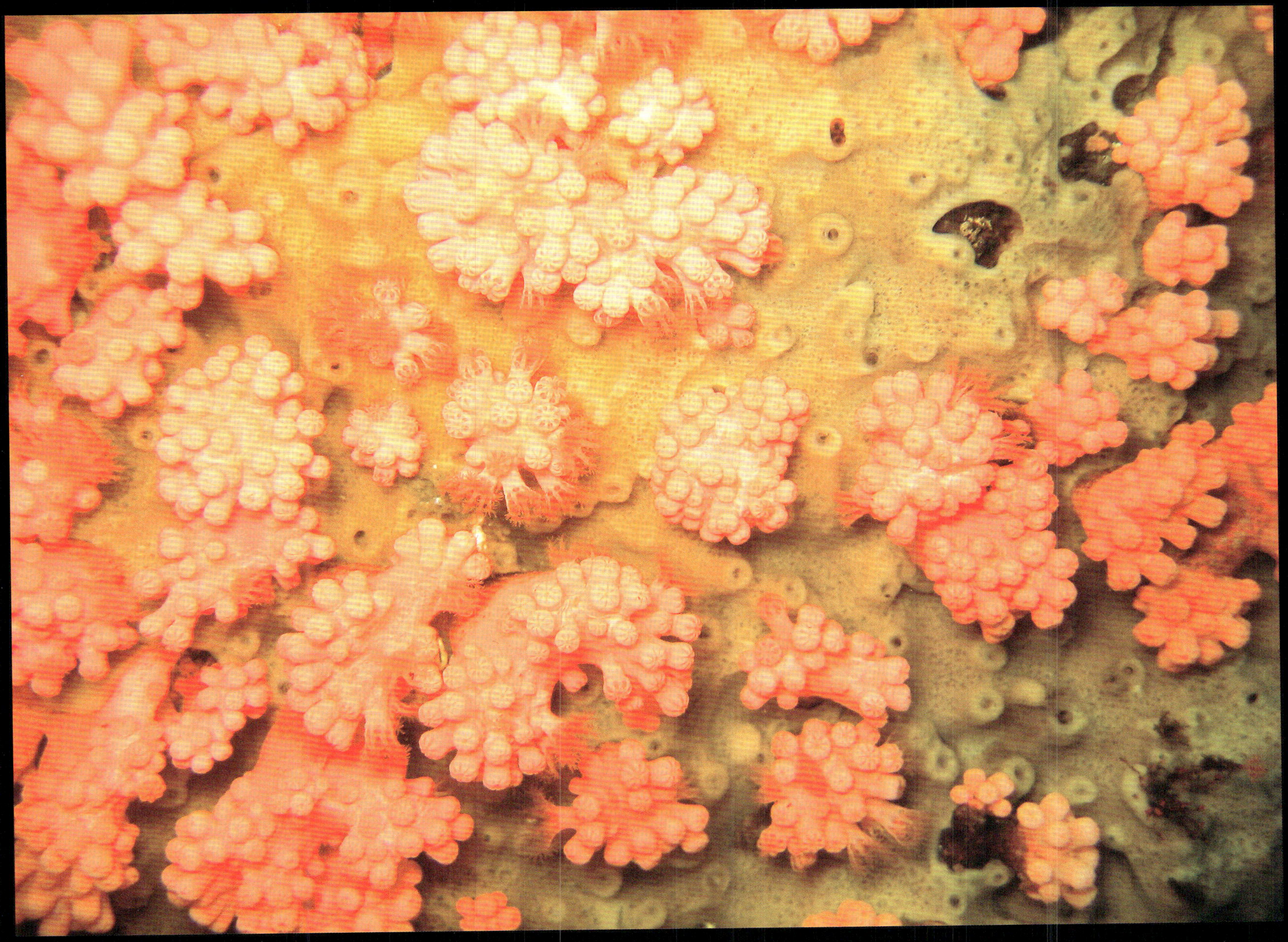

Pink soft corals, such as *Gersemia rubiformis*, are a beautiful sight in the Aleutian Islands. Soft corals, like their cousins the sea anemones, are usually found in high-current passages where they can maximize their exposure to passing food. The soft coral patches typically do not exceed 8 in wide (20 cm). *Attu Island, H. Chenelot photo.*

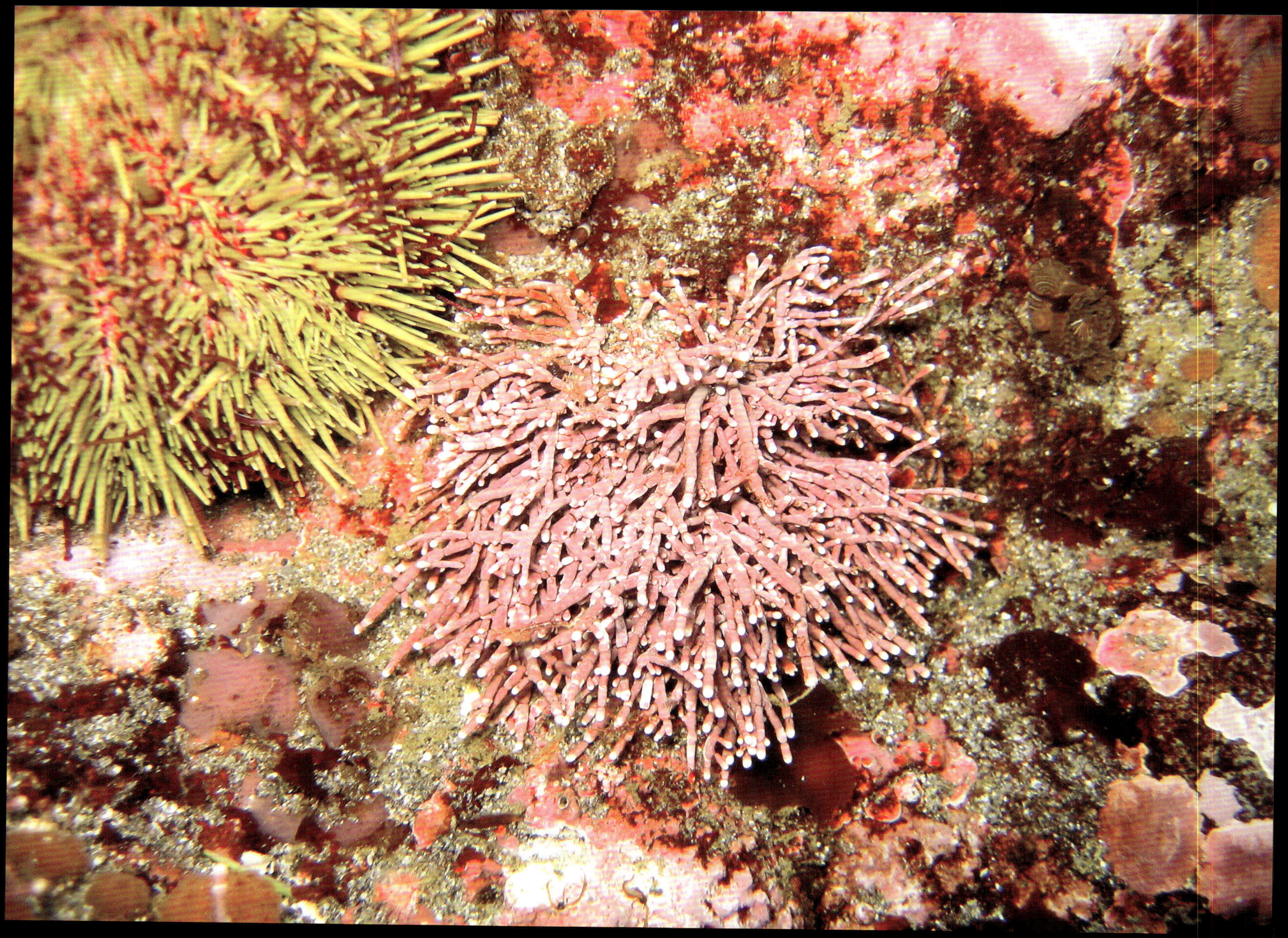

The chalky coral seaweed (*Pachyarthron cretaceum*) is a coralline alga. Its calcified branches remain somewhat flexible thanks to the presence of uncalcified joints, called genicula, between the hardened segments. This species is found from Russia and Japan, throughout the Aleutians, to northern Washington. *Shemya Island, H. Chenelot photo.*

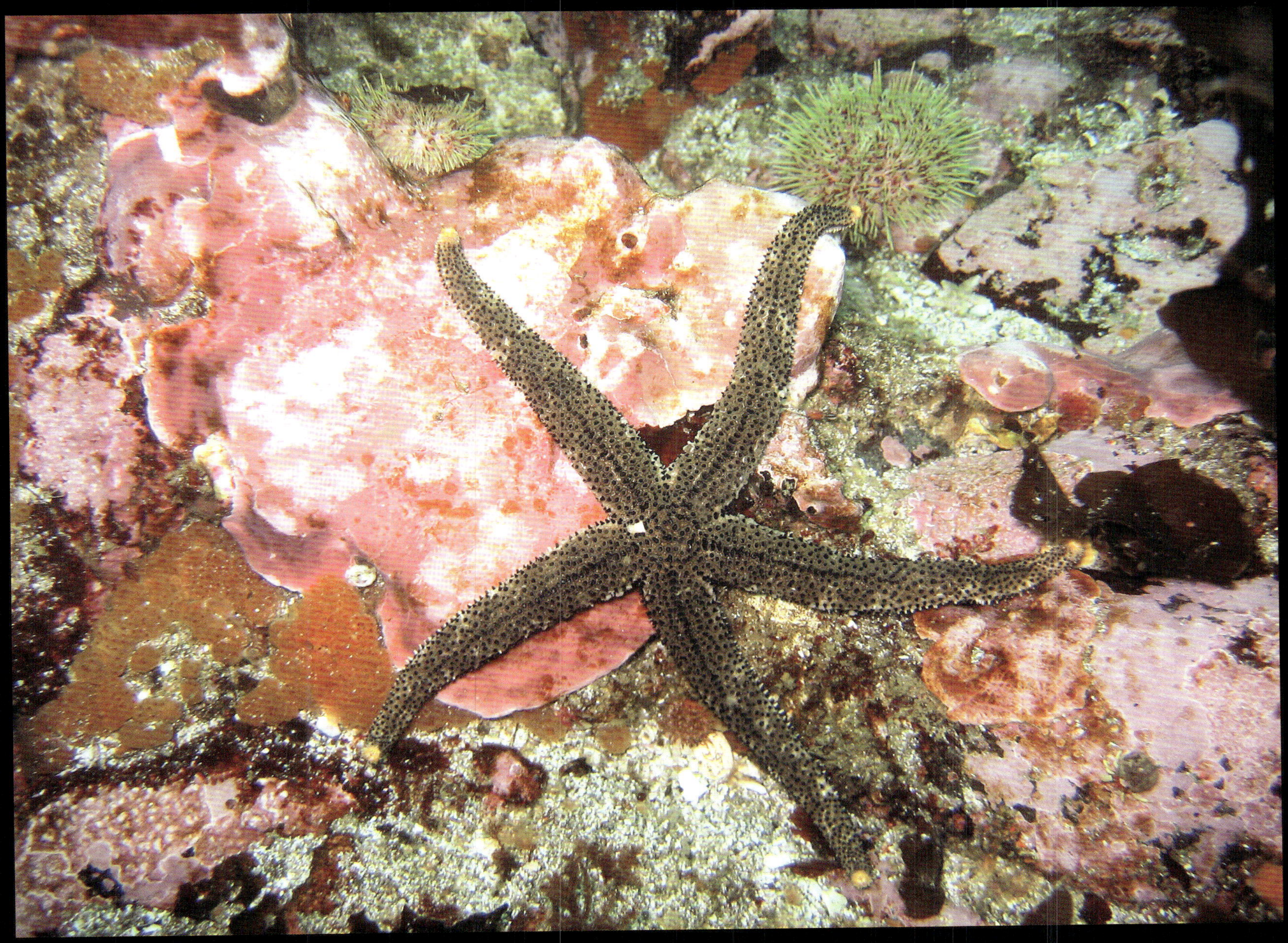

The black-spined star (*Lethasterias nanimensis*) is a large (to 24 in, 60 cm diameter) star that occurs throughout the Aleutians. Its black-tipped spines set it apart from other sea stars. It feeds on invertebrates such as snails, cockles, sand dollars, and sea cucumbers, on soft and hard surfaces. *Massacre Bay, Murder Point, Attu Island, H. Chenelot photo.*

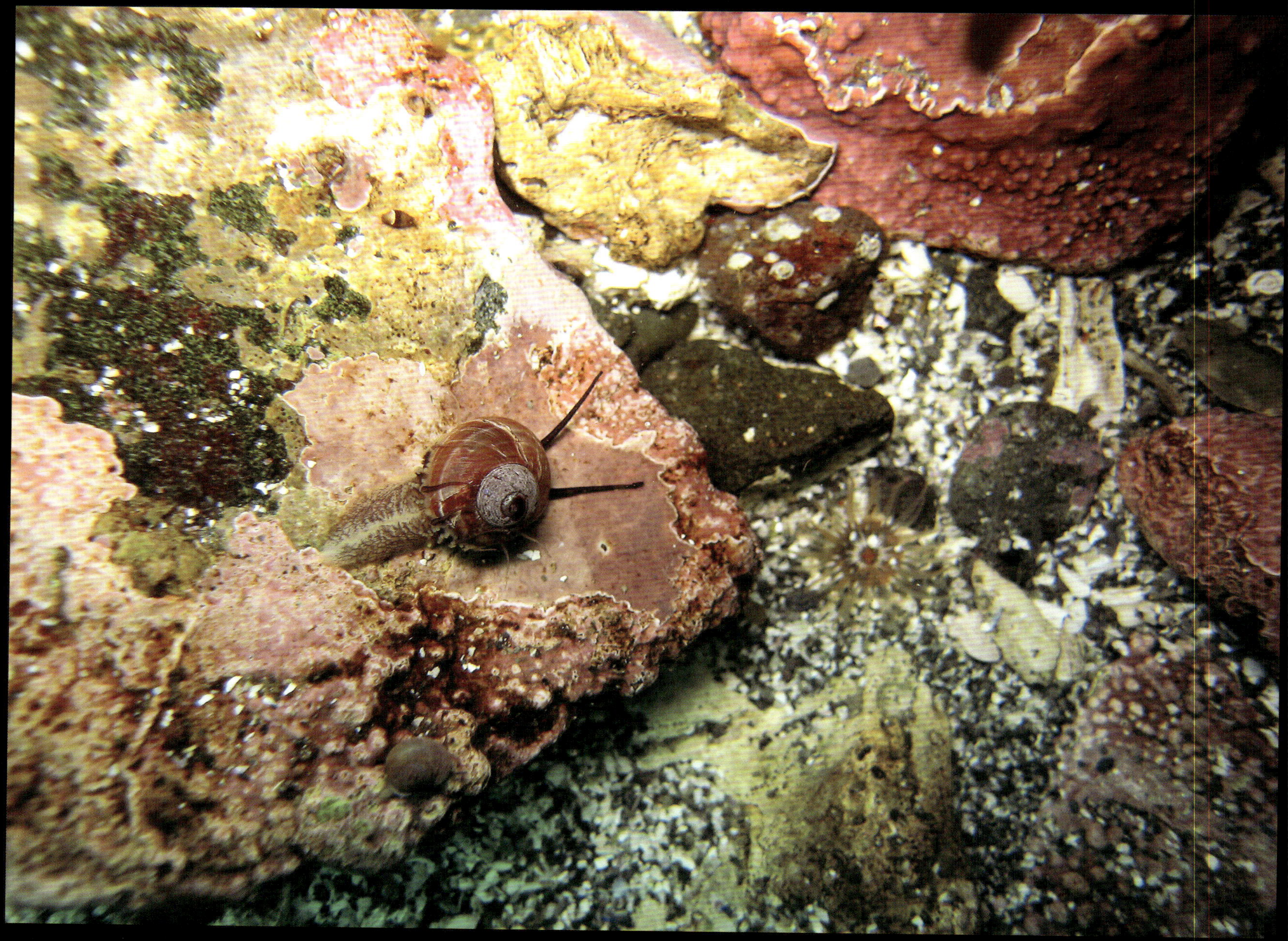

The vortex margarite snail (*Margarites vorticiferus*) has a distinctive depressed (flattened) shell. It is widely distributed from Pacific arctic seas and Japan to San Pedro, California, and is common in shallow waters to 360 ft (110 m). *Massacre Bay, Murder Point, Attu Island, S. Jewett photo.*

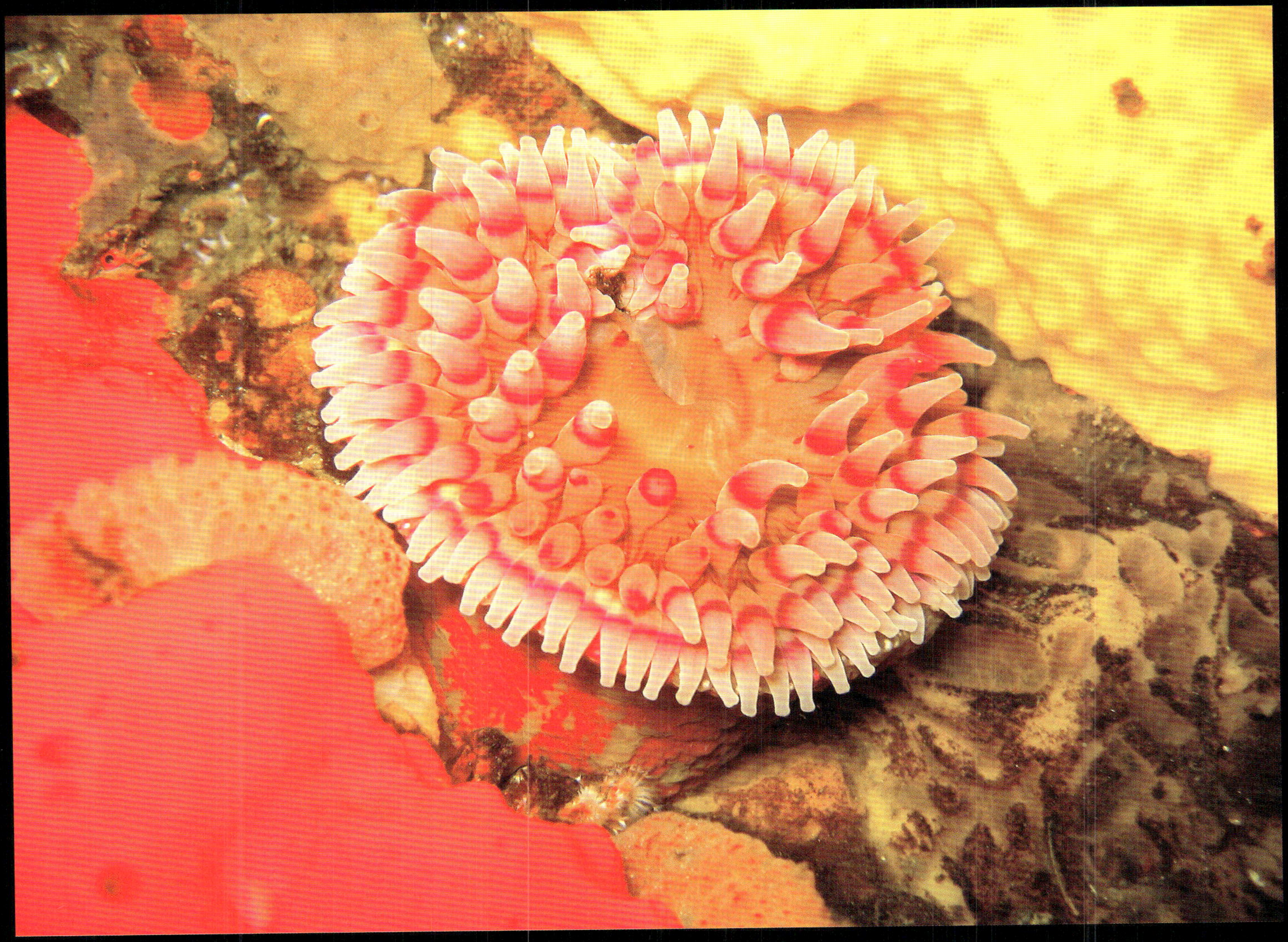

The painted anemone (*Urticina grebelnyi*), with its many color variations, is common in the Aleutian Islands. This species is sometimes called the Christmas tree anemone because a common color morph has green and red blotches along the column, as in this photo. The tentacles, used to capture passing prey, are long and banded. *Attu Island, H. Chenelot photo.*

Literature Resources

Bacon, C.E., W.M. Jarman, J.A. Estes, M. Simon, and R.J. Norstrom. 1999. Comparison of organochlorine contaminants among sea otter (*Enhydra lutris*) populations in California and Alaska. Environmental Toxicology and Chemistry 18:452-458.

Barron, M.G., R. Heintz, and M.M. Krahn. 2003. Contaminant exposure and effects in pinnipeds: Implications for Steller sea lion declines in Alaska. Science of the Total Environment 311:111-133. doi:10.1016/S0048-9697(03)00140-2

Burger, J., M. Gochfeld, and S.C. Jewett. 2007. Radionuclides in benthic invertebrates from Amchitka and Kiska islands in the Aleutian chain, Alaska. Environmental Monitoring and Assessment 128:329-341. doi:10.1007/s10661-006-9316-4

Burger, J., M. Gochfeld, D. Kosson, C.W. Powers, S. Jewett, B. Friedlander, H. Chenelot, C.D. Volz, and C. Jeitner. 2006. Radionuclides in marine macroalgae from Amchitka and Kiska islands in the Aleutians: Establishing a baseline for future biomonitoring. Journal of Environmental Radioactivity 91:27-40. doi:10.1016/j.jenvrad.2006.08.003

Burger, J., M. Gochfeld, D. Kosson, C.W. Powers, B. Friedlander, M. Stabin, D. Favret, S.C. Jewett, D. Snigaroff, R. Snigaroff, T. Stamm, J. Weston, and C. Jeitner. 2007. Radionuclides in marine fishes and birds from Amchitka and Kiska islands in the Aleutians: Establishing a baseline. Health Physics 92(3):265-279. doi:10.1097/01.HP.0000248123.27888.d0

Chenelot, H., S. Jewett, and M. Hoberg. 2008. Invertebrate communities associated with various substrates in the nearshore eastern Aleutian Islands, with emphasis on thick crustose coralline algae. In: P. Bruggeman and N.W. Pollock (eds.), Diving for science 2008. Proceedings of the 27th American Academy of Underwater Sciences Symposium, Dauphin Island, Alabama. AAUS, pp. 13-36. doi:10.1007/s12526-010-0071-y

Chenelot, H., S.C. Jewett, and M.K. Hoberg. In press. Macrobenthos of the nearshore Aleutian archipelago, with emphasis on invertebrates associated with *Clathromorphum nereostratum* (Rhodophyta, Corallinaceae). Marine Biodiversity.

Clark, R.N., and S.C. Jewett. 2010. A new genus and thirteen new species of sea stars (Asteroidea: Echinasteridae) from the Aleutian Island archipelago. Zootaxa 2571:1-36.

Clark, R.N., and S.C. Jewett. In press. A new species of sea star, *Hippasteria* (Asteroidea: Goniasteridae), from the Aleutian Islands. Zootaxa.

Eash-Loucks, W., S.C. Jewett, D.G. Fautin, M.K. Hoberg, and H. Chenelot. 2010. *Ptychodactis aleutiensis*, a new species of ptychodactiarian sea anemone (Cnidaria: Anthozoa: Actiniaria) from the Aleutian Islands, Alaska. Marine Biology Research 6:570-578. doi:10.1080/17451000903476933

Estes, J.A., and J.F. Palmisano. 1974. Sea otters: Their role in structuring nearshore communities. Science 185:1058-1060. doi:10.1126/science.185.4156.1058

Estes, J.A., and D.O. Duggins. 1995. Sea otters and kelp forests in Alaska: Generality and variation in a community ecological paradigm. Ecological Monographs 65:75-100. doi:10.2307/2937159

Estes, J.A., M.T. Tinker, and J.L. Bodkin. 2010. Using ecological function to develop recovery criteria for depleted species: Sea otters and kelp forests in the Aleutian archipelago. Conservation Biology. doi:10.1111/j.1523-1739.2009.01428.x

Estes, J.A., C.E. Bacon, W.M. Jarman, R.J. Norstrom, R.G. Anthony, and A.K. Miles. 1997. Organochlorines in sea otters and bald eagles from the Aleutian archipelago. Marine Pollution Bulletin 34:486-490. doi:10.1016/S0025-326X(96)00178-6

Jewett, S.C., J. Bodkin, H. Chenelot, G. Esslinger, and M. Hoberg. 2010. The nearshore benthic community of Kasatochi Island, one year after the 2008 volcanic eruption. Arctic, Antarctic, and Alpine Research 42:315-324. doi:10.1657/1938-4246-42.3.315

Jewett, S.C., M. Hoberg, H. Chenelot, S. Harper, J. Burger, and M. Gochfeld. 2006. Scuba techniques used in risk assessment of possible nuclear leakage around Amchitka Island, Alaska. In: J.M. Godfrey and S.E. Shumway (eds.), Diving for science 2005: Proceedings of the American Academy of Underwater Sciences 24th Annual Diving Symposium. University of Connecticut Sea Grant College Program, CTSG-06-03, pp. 143-156.

Jewett, S.C., R. Brewer, H. Chenelot, R. Clark, D. Dasher, S. Harper, and M. Hoberg. 2008. Scuba techniques for the Alaska Monitoring and Assessment Program (AKMAP) of the Aleutian Islands, Alaska. In: P. Bruggeman and N.W. Pollock (eds.), Diving for science 2008. Proceedings of the 27th American Academy of Underwater Sciences Symposium, Dauphin Island, Alabama. AAUS, pp. 71-89.

Johnson, T. 2003. The Bering Sea and Aleutian Islands: Region of wonders. Alaska Sea Grant College Program, University of Alaska Fairbanks. 199 pp.

Kawai, H., T. Hanyuda, M. Lindeberg, and S.C. Lindstrom. 2008. Morphology and molecular phylogeny of *Aureophycus aleuticus* gen. et sp. nov. (Laminariales, Phaeophyceae) from the Aleutian Islands. Journal of Phycology 44:1013-1021. doi:10.1111/j.1529-8817.2008.00548.x

Konar, B. 2000. Limited effects of a keystone species: Trends of sea otters and kelp forests at the Semichi Islands, Alaska. Marine Ecology Progress Series 199:271-280. doi:10.3354/meps199271

Konar, B., and J.A. Estes. 2003. The stability of boundary regions between kelp beds and deforested areas. Ecology 84:174-185. doi:10.1890/0012-9658(2003)084[0174:TSOBRB]2.0.CO;2

Lamb, A., and B.P. Hanby. 2005. Marine life of the Pacific Northwest: A photographic encyclopedia. Harbour Publishing, Madeira Park, B.C., Canada. 398 pp.

Lindeberg, M.R., and S.C. Lindstrom. 2010. Field guide to seaweeds of Alaska. Alaska Sea Grant College Program, University of Alaska Fairbanks. 192 pp.

Merritt, M.L., and R.G. Fuller. 1977. The environment of Amchitka Island, Alaska. U.S. Energy Research and Development Association, Technical Information Center, TID-26712. 682 pp.

Miles, A.K., M.A. Ricca, R.G. Anthony, and J.A. Estes. 2009. Organochlorine contaminants in fishes from coastal waters of Amukta Pass, Aleutian Islands, Alaska, USA. Environmental Toxicology and Chemistry 28(8):1643-1654. doi:10.1897/08-498.1

Morgan, L. (ed.). 1980. The Aleutians. Alaska Geographic 7(3). Alaska Geographic Society, Anchorage.

Morris, R. 2006. Overview of M/V *Selendang Ayu* response. In: R. Brewer (ed.), The *Selendang Ayu* oil spill: Lessons learned. Alaska Sea Grant College Program, University of Alaska Fairbanks, pp. 5-16.

Reisewitz, S.E., J.A. Estes, and C.A. Simenstad. 2006. Indirect food web interactions: Sea otters and kelp forest fishes in the Aleutian archipelago. Oecologia 146:623-631. doi:10.1007/s00442-005-0230-1

Rennick, P. (ed.). 1995. The Aleutian Islands. Alaska Geographic 22(2). Alaska Geographic Society, Anchorage.

Rennick, P. (ed.). 1995. World War II in Alaska. Alaska Geographic 22(4). Alaska Geographic Society, Anchorage.

Sanamyan, K., and N. Sanamyan. 2007. The underwater world of Kamchatka: Cruise around Avacha Bay. VeGa Publishing, Moscow, Russia. 256 pp.

Schumacher, J.D., G.H. Kruse, and S.A. Macklin (eds.). 2005. The Aleutian ecosystem: Processes controlling variability in productivity and ecosystem structure. Fisheries Oceanography 14 (Suppl. 1):1-306. doi:10.1111/j.1365-2419.2005.00376.x

Springer, A.M., C.P. McRoy, and M.V. Flint. 1996. The Bering Sea green belt: Shelf-edge processes and ecosystem production. Fisheries Oceanography 5:205-233. doi:10.1111/j.1365-2419.1996.tb00118.x

Witherell, D., and D. Woodby. 2005. Application of marine protected areas for sustainable production and marine biodiversity off Alaska. Marine Fisheries Review 67(1):1-27.

Acknowledgments

The 2004 research cruise was funded by the Consortium for Risk Evaluation with Stakeholder Participation (CRESP) through the U.S. Department of Energy to the University of Alaska Fairbanks (S. Jewett). We thank the 2004 crew of the F/V *Ocean Explorer* (Captain Ray Haddon) for making our fieldwork possible and pleasant. Special thanks go to Joanna Burger and Michael Gochfeld for their assistance with many phases of the CRESP project. Thanks also to the outstanding editorial staff at the Alaska Sea Grant College Program whose expertise in book publishing made this book a reality, and thanks to Doug Schneider for writing the introduction.

The 2006-2007 research cruises were funded by the U.S. Environmental Protection Agency, Office of Research and Development through the Alaska Department of Environmental Conservation and the University of Alaska Fairbanks (S. Jewett). We appreciate the help of the crew of the R/V *Norseman* (Captain Paul Tate). We thank the following individuals for their support in the research cruises: Nicholas Dallman, Douglas Dasher, James Gendron, Terri Lomax (ADEC), and Roger Deffendall (volunteer diver, Unalaska, Alaska). We also thank the crew of the R/V *Thompson* and UAF principal investigator, Dr. Brenda Konar, for the opportunity to dive and photograph in 2008 (H. Chenelot).

We thank the following for their taxonomic assistance: William Austin (sponges); Roger Clark (mollusks and sea stars); Meg Daly (sea anemones); Daphne Fautin (sea anemones); Max K. Hoberg (benthic invertebrates); Gretchen Lambert (tunicates); Philip Lambert (echinoderms); Mandy Lindeberg (algae); Catherine Mecklenburg (fishes); and Karen Sanamyan (tunicates). Carol Ladd, National Oceanic and Atmospheric Administration, provided the figure on Aleutian passes and Karen Peterson, U.S. Geological Survey, led us to the base map of Alaska. We also appreciate input for the dedication from Bob King, Alaska fisheries historian and advisor to U.S. Senator Mark Begich, and from Arliss Sturgulewski, Vera Alexander, and Michael Castellini.

In addition to the photographs taken by the authors, the following individuals kindly allowed us to use theirs: Joanna Burger, Roger N. Clark, Nicholas Dallman, Chris Ford, Max K. Hoberg, Mandy Lindeberg, Jerry Morris, and Chris Waythomas. Photo of Ted Stevens on page vi by Kurt Byers, and photo of Sylvia Earle on page viii © Kip Evans Photography.

Partial funding for publishing was made available by the City of Unalaska, Alaska; the Ounalashka Corporation; Aleut Corporation; Norseman Maritime Charters LLC; the Aleutian Pribilof Island Association, the Rasmuson Foundation, and the Alaska Sea Grant College Program.

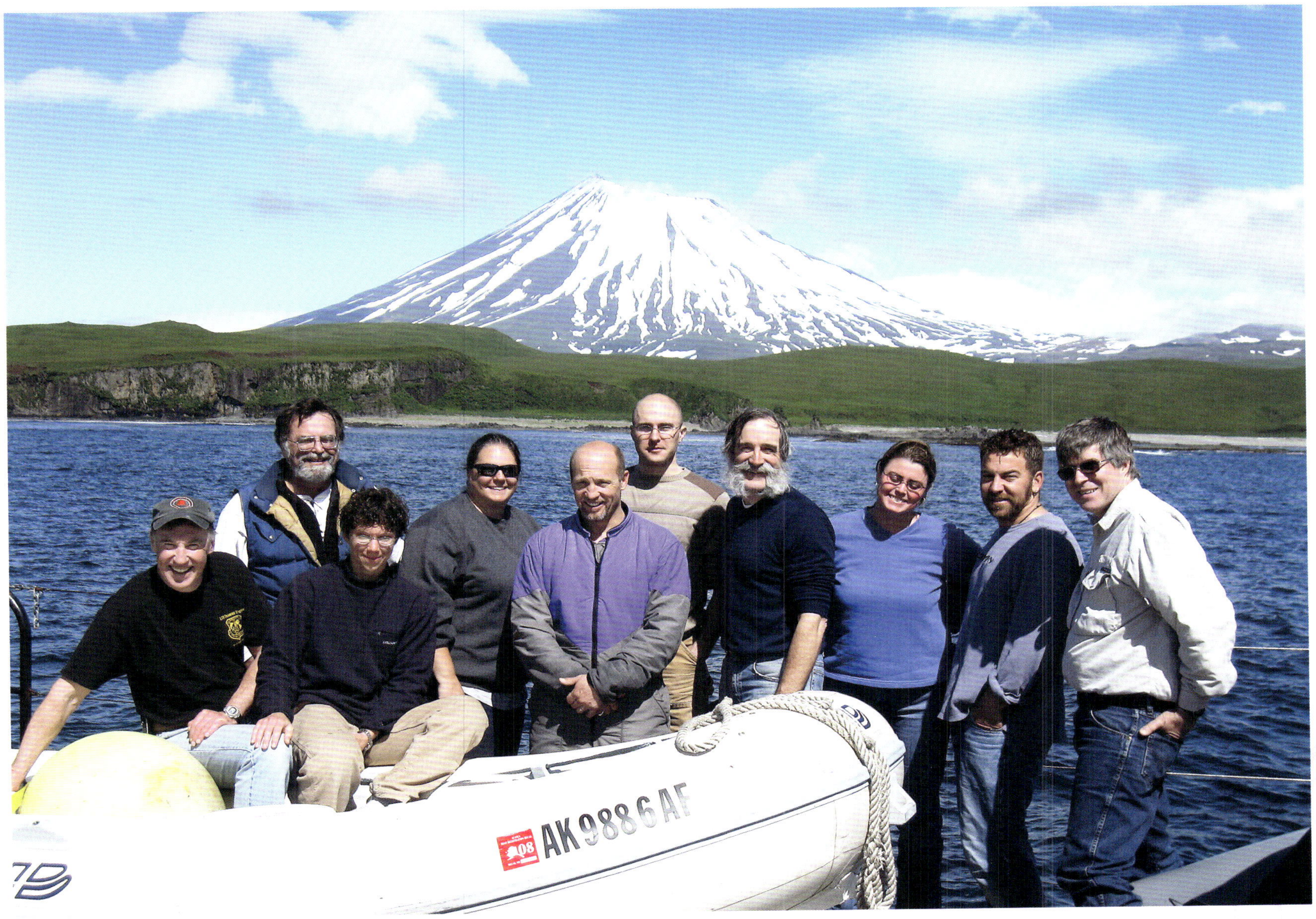

Research expedition science crew in 2006-2007 (left to right): Stephen Jewett, James Gendron, Héloïse Chenelot, Mandy Lindeberg, Roger Clark, Shawn Harper, Max Hoberg, Terri Lomax, Reid Brewer, and Doug Dasher. Mount Vsevidof on Umnak Island is in the background.

Index

A

B

C

D

E

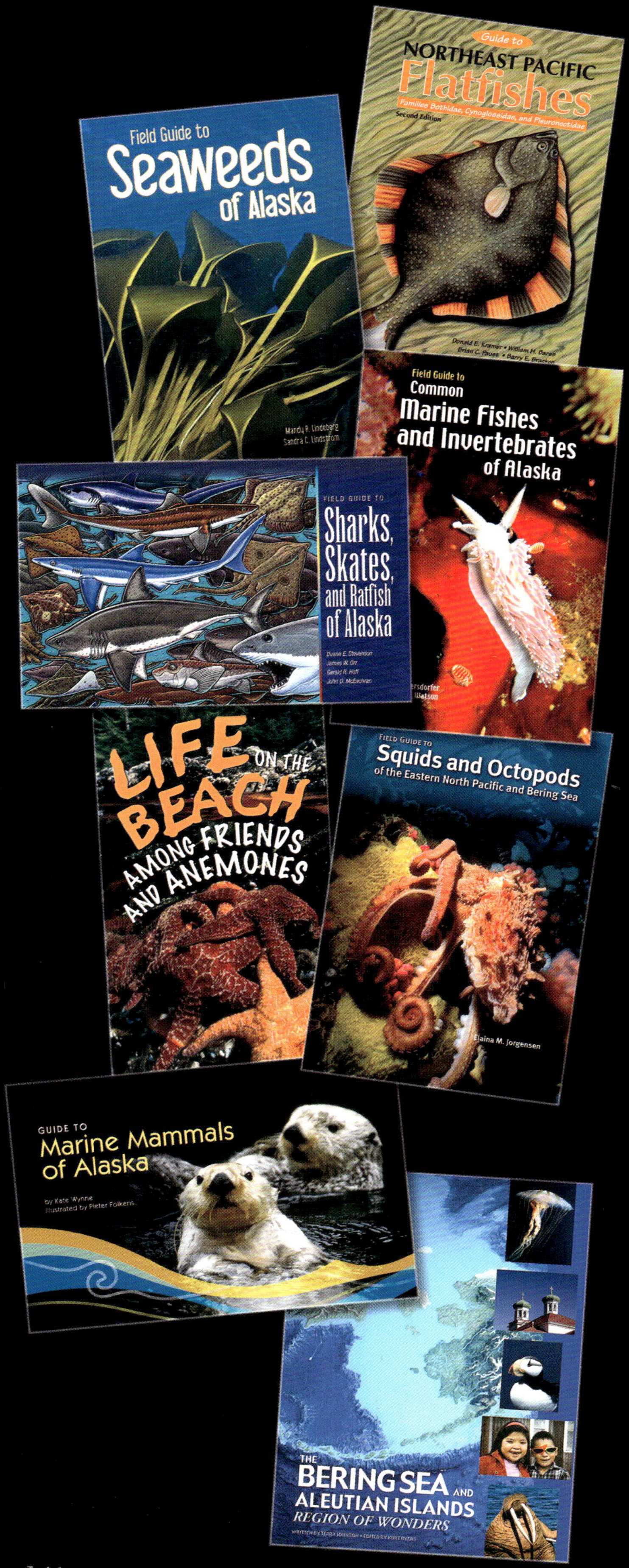

Alaska's Seas and Coasts

Learn much more about Alaska's amazing seas and coasts with other Alaska Sea Grant College Program books and videos, including:

- The Bering Sea and Aleutian Islands: Region of Wonders
- Field Guide to Common Marine Fishes and Invertebrates of Alaska (PDF)
- Field Guide to Marine Mammals of Alaska
- Guide to Northeast Pacific Flatfishes
- Field Guide to Squids and Octopods of the Eastern North Pacific and Bering Sea
- Field Guide to Sharks, Skates, and Ratfish of Alaska
- Field Guide to Bird Nests and Eggs of Alaska's Coastal Tundra
- Field Guide to Seaweeds of Alaska
- Common Edible Seaweeds in the Gulf of Alaska
- Imam Cimiucia: Our Changing Sea
- Ocean Treasure: Commercial Fishing in Alaska
- Life on the Beach: Among Friends and Anemones (children's video)

These are just some of the publications and videos available from Alaska Sea Grant. Call toll-free for a free catalog, 1-888-789-0090, or visit the Alaska Sea Grant bookstore: **WWW.ALASKASEAGRANT.ORG**

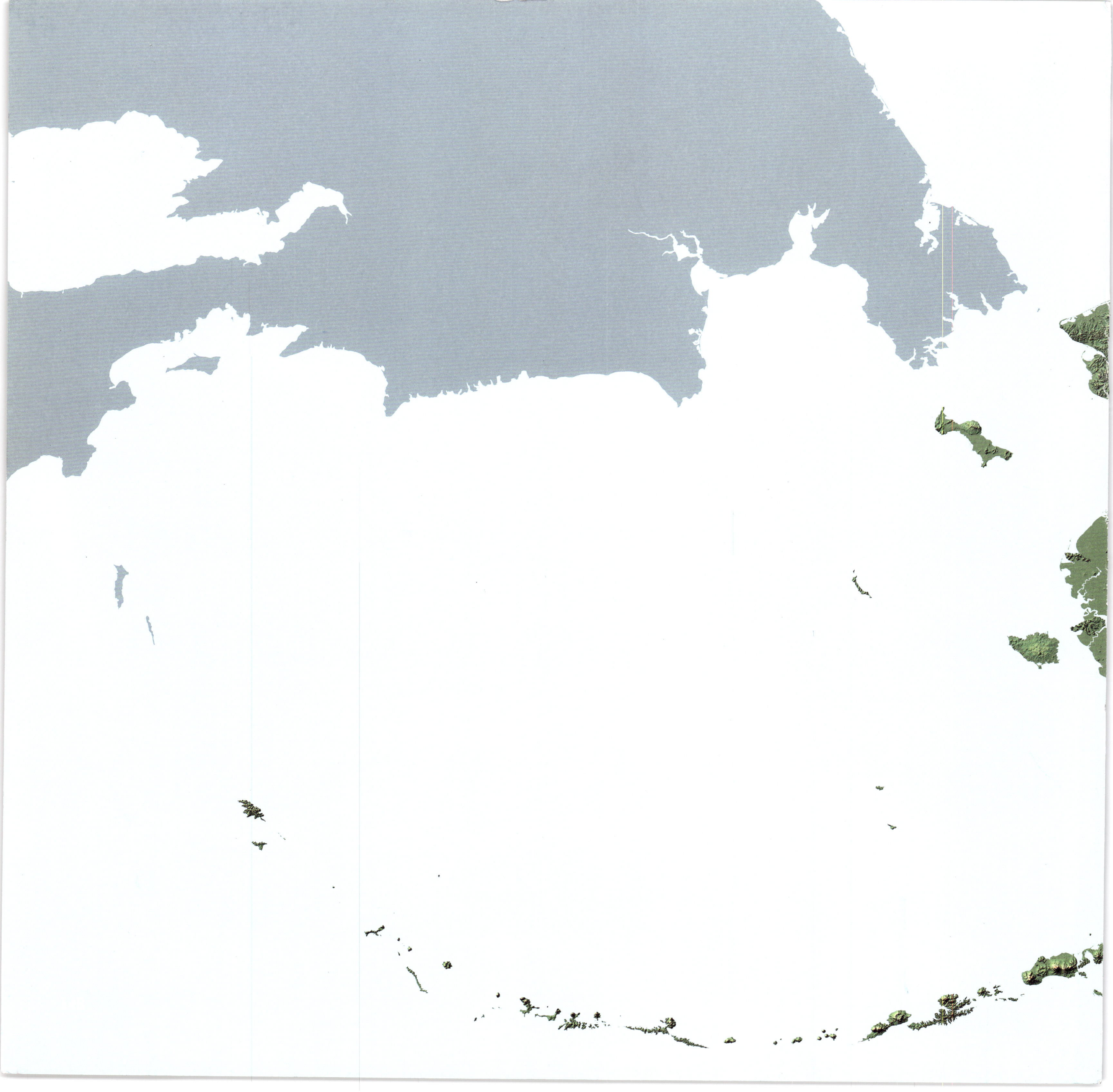